100
YEARS
SIMON &
SCHUSTER

THE AIR THEY BREATHE

A Pediatrician on the Frontlines of Climate Change

DEBRA HENDRICKSON, MD

SIMON & SCHUSTER

New York London Toronto Sydney New Delhi

1230 Avenue of the Americas
New York, NY 10020

Some names have been changed.

First Simon & Schuster hardcover edition July 2024

Interior design by Wendy Blum

Manufactured in the United States of America

1 3 5 7 9 10 8 6 4 2

Library of Congress Cataloging-in-Publication Data has been applied for.

ISBN 978-1-5011-9713-0
ISBN 978-1-5011-9715-4 (ebook)

CONTENTS

For my children

Without showing what happened to a child, the world cannot understand.

—Keiko Ogura

Introduction

HEROES

In a musty box from my childhood sits a collection of 12-cent *Superman* comic books. When I was in grade school I would trek to the local grocery store every month to buy them, my weekly allowance in my pocket and my dog Nikki in tow. The store was owned in those days by a family in our town, and the clerk, a lady in her fifties with cat-eye glasses and a warm smile, knew me and every other kid in the area; she would always ask if my money was burning a hole in my pocket and that was why I had to spend it so fast.

The comics are fragile now, ink fading, pages yellow and thin, but I remember their newness so clearly—how I rushed with the latest issue to meet neighborhood kids at the marsh in our small California town, sharing Superman's heroics while eating beef jerky and fishing for crabs. We would reenact our favorite stories with gusto, pinning towels to our shoulders as capes and leaping from rocks in flight. (The boys always wanted me to play Lois Lane, but I never went along.) By the time I got home the comic book was mud-spattered, worn, and memorized, cover to cover.

We all knew this about Superman: he became a hero through tragedy. His scientist father tried valiantly to convince the Krypton council that everyone on their world was in mortal danger, and needed to be saved from a doomed planet. It was no use: he was ridiculed, his warning ignored. To save their son, Superman's parents put him on a spaceship to Earth. Krypton exploded into millions of pieces.

I could never have imagined, as a child, that in my lifetime a parallel story would unfold on Earth. Dying worlds were for fiction; saving humanity from doom was a game. We took the permanence of our own planet—beautiful, enormous, and filled with life—for granted. We did not stop to think, when playing with the marsh's water bugs and snakes, polliwogs and herons, that each of these creatures is a miracle, or that our planet itself might be unique in all the universe. Who can imagine the end of the world?

Nor did we notice, as children, that we, too, were creatures of the marsh, tied in countless unseen ways to the place we lived. Our legs waded in cool, murky water; our hands scooped small life into jars to carry home. Our eyes watched clouds flow over us, while our lungs took in the clean air.

But the marsh, full of wonder, held darker clues of what was to come. It behaved like a living thing, sometimes placid, at other times lashing out, its streams rising into backyards, roads, and houses. One Christmas my sister and I woke to find our presents floating in a foot of water in the living room; our street had suddenly become a river. The more houses and roads built in and near the marsh, the worse the flooding; development was overwhelming its ability to absorb runoff from uphill.

I think of that today because the Earth, though massive and unalterable in our minds, clearly has its limits. For decades, scientists have issued increasingly dire warnings about the dangers of polluting the atmosphere with greenhouse gases such as carbon dioxide and methane. These gases, produced mainly by the burning of fossil fuels, trap heat from the sun that would otherwise radiate back to space. Our planet's atmosphere can absorb only so many tons of these pollutants before the complex yet stable pattern of precipitation and temperature that sustains life—our climate—becomes unstable and dangerous.

Just as in the comic myth, there has been some resistance to these facts. But denial of reality hasn't changed it. Global temperatures rise at an alarming pace; each decade is warmer than the last. The eight years prior to 2023 were the hottest ever recorded, fueling natural disasters of unprecedented power and destruction. Then 2023 shattered Earth's heat records once again. This relentless warming trend would quickly flatten if we stopped polluting the atmosphere with carbon; the problem and its solutions are well understood. Yet staggering volumes of fossil fuel emissions continue. Millions of lives, whole species, and entire countries are at risk from extreme weather and rising seas. We are on the verge of rendering our beautiful world a hostile, less habitable place.

Children, with their lives ahead of them, have the most at stake—and they know it. Surveys show that a large majority of children, from school-age to teens, worry about the changing climate. They rightly see it as a threat to their own well-being, casting a shadow over their dreams. But the percentage who believe that adults will address the crisis, and save the Earth, is heartbreakingly small. Our kids understand: this is not pretend. Their planet is in trouble.

In the summer of 2018, a sixteen-year-old told me during a checkup, when I asked about her plans after high school, that she hoped to be a science teacher. "That's great," I began, when she politely cut me off: she wanted to work with kids because she was questioning whether to have any of her own. A school project on global warming, which included interviewing a climate scientist at the local university, had been keeping her awake. "I'm not sure it would be right," she said, "to bring kids into a world like that."

She glanced up for my reaction as she said it. I was sitting across from her on my wheeled stool, with her chart open on my lap; I had just written "loves soccer" and "future plans—" then stopped. I noticed her hands gripping the edge of the exam table, as if it might suddenly move. Her legs were crossed, swinging a little; short, light-brown hair was looped neatly behind one ear. She was young, and healthy, with purple-framed eyeglasses and flip-flops dangling from her feet. A tear suddenly fell down her cheek.

And I struggled with what to say.

Wildfires, hurricanes, and heat waves make headlines. Small moments with my patients never will. But what is happening in my clinic tells another story of this strange and unsettling time. Children are bearing the weight, in their lungs and hearts and minds, of our madness.

I am a pediatrician in the fastest-warming city in the United States. Reno lies four hours' drive east of my childhood marsh, within sight of the Nevada-California border. It spans a small river valley and a series of sagebrush hills, joining the Great Basin Desert and Sierra Nevada mountains like pieces of a massive, continental puzzle. Year by year, through the windows of the hospital and clinic where I work, I watch our sky and forests changing as the thermometer climbs. Inside those buildings, I care for the children of my town.

And from that vantage point I can see that a marsh and the children playing in it are not separate; they are part of the same fabric, and their fates are intertwined. The climate crisis is a health crisis, and it is a health crisis, first and foremost, for children.

The American Academy of Pediatrics has warned about climate threats to children for over a decade; almost 90 percent of global disease attributable to the crisis is occurring in those under five years old. Most parents I talk to, every day—who lovingly buckle their children into car seats, brush their teeth, and bring them in for checkups and vaccines—don't know this. Nor do they realize that their children are already being affected by the changes we have set in motion.

That's because the health problems linked to climate change, much like the weather disasters it spawns, seem like familiar events. Just as hurricanes, wildfires, and floods have always been part of human history, doctors have treated the illnesses associated with global warming—asthma and allergies, heatstroke, trauma, infections, and malnutrition—for centuries. What is changing is the intensity, frequency, and prevalence of problems we have always known.

I see it in my clinic, like the edge of a flood lapping beneath the door. When a mother says that her daughter suffers with hay fever most of the year now, instead of just in spring. Or a father questions why his son fainted from the heat during football practice in late September, when our weather usually cools. When a child struggles to pay attention in a new school after her family fled the destruction of a record-breaking hurricane and moved cross-country. Or when I treat more tick bites in a single summer than I have in the previous decade.

While everyone is harmed by the shifts taking place around us (parents and doctors cough in smoky air, too), in no other age group does the environment have as much power to shape a person's mental and physical health. From the moment of conception until adulthood, a child's body and mind are forming, every day. Like wet clay on a potter's wheel, the lightest touch can have profound effect; yet as the clay sets—as our major organs finish growing and developing—the same touch leaves little or no mark. Children are also more susceptible because of their size; just as an adult dose of medication would harm a child, the same "dose" of air pollution or heat is more dangerous for a small body than an adult's.

But one aspect of this crisis is particularly confined to youth, and stays with me each night as I turn off the lights and head home. Our children are planning their lives in circumstances no prior generation has seen. Childhood has always been a risky period of life; throughout history, babies and children have met peril, from polio to famine, from cyclones to war. Yet they have never had to face, in quite this way, the potential loss of the future itself.

I set down the girl's chart and handed her a tissue. In the hallway my medical assistant asked *Which ear hurts, buddy?* as the next exam room door shut. We sat for a moment; then I asked if her mother, also in the hallway, knew how she felt. She wiped her nose and stared at the Kleenex in her hands. "I don't like to say anything," she said, "because she just gets this look on her face."

"What kind of look?"

She shrugged. "Like she's scared. And sad."

"So you don't want to worry her."

She shook her head no.

It's our job to worry about you, I thought, but didn't say. Because what defense would there be, if that were spoken.

I wanted to pin a towel to her shoulders and tell her she could save the Earth. I wanted her to understand that despair was like kryptonite in the hands of powerful men: a weapon wielded to get their way. I wanted her to feel some of what I felt, welling in my throat.

Because our warming world is not just a health and existential threat; it is the greatest moral crisis humanity has ever faced. This young girl was not the victim of an accident. All the children alive today, and all who will ever live, were chosen to bear this burden by the men who ruled the world.

Those men led the industry that had laid our path to this moment, years ago; their decisions were now here, in my exam room. In 1977, when I was about this girl's age, Exxon scientist James F. Black realized that carbon dioxide in the atmosphere was rising, and the planet was warming, due to the burning of fossil fuels. He presented a startling warning to the company: within decades, continued use of its products would radically alter the Earth.

The Exxon executives who heard his report were alarmed. They funded millions of dollars of research, all confirming his ominous findings. But by 1988, when NASA scientist James Hansen told Congress the same thing, the heads of the largest oil and gas companies had made a choice. And that choice would shape the future of billions of people, living and unborn.

Their minutes and memos did not mention my patient. They never thought about the life of one young girl in my clinic. But I thought of them now, as I sat with her: powerful people we would never meet, deciding her fate. Men who risked the world, and made a child pay their debt.

Outside the windows of the medical center that afternoon, Reno was in the midst of the hottest month in its history, breaking a record set just the year before. In the Sierras and beyond, heat was baking California's forests, where millions of trees had been killed by drought and bark beetles, whose numbers were exploding in the warming climate. Any day now, a spark from a power line, or a tire rim scraping on asphalt, would ignite that tinder. And smoke would pour over the mountains again, as it does every summer, from the fires consuming places I once knew.

And that smoke would be carried in the air that she, and all my other patients, pulled into their bodies; it would irritate their eyes and chests and moods, and the smiles I usually saw from well children would be replaced, as the weeks wore on, by flat expressions. The smoke would even snake through the hospital's filters to the newborn nursery, where I would suddenly smell it as I turned a baby over onto my forearm, gently, to check his back. When I returned him to his bassinette and looked at his sleeping face, I would wonder what remnants of other people's lives he was breathing on his first day.

As children left our building with their parents that month, they would have to contend not just with smoke, but with the ongoing heat wave, which would *whoosh* through the sliding glass doors like a giant's hand. Its grip would trigger their skin glands to sweat, blood vessels to dilate, and pulses to quicken. At night, it would trouble their sleep—because evening temperatures no longer drop like they once did, here in our high desert climate. And I would have little to offer their worried parents, except a warning to keep children indoors—the opposite of my usual advice. Few needed to be told. We could all see that the world outside had become alien to us, and unsafe.

I didn't know that any of this was in our future as I sat with that teenage girl in my clinic. I didn't know that shortly after the Carr and Mendocino Complex fires, another spark, just west of here, would take the town of Paradise; that it would send refugees—grandparents, aunts, uncles, and cousins—into our city, where they would double up in the bedrooms of my patients. But I did know; we all knew. And I thought of her as it happened.

The air that summer would eventually cool, and the city and mountains would reappear from haze; we would go back to forgetting the crisis we are in, and the foreshadowing we had been given. Months later, a few drops of rain would fall, and children would run outside to catch it on their tongues. And we would look at them, and this world we know, and tell ourselves nothing was changing.

I would be passing through a glass walkway on the day of that first rain, going from my office building to the pediatric floor, when I saw these children outside, playing in the hospital garden. Among them was a six-year-old I knew. He had just left my clinic with his mother and toddler brother, and was now standing in the garden below, face upturned and mouth open.

I paused to watch him. Every minute or so he would jump up and down, laughing and wiping his face with his palms, because a raindrop had hit him on the nose or cheek. As other staff hurried past me in the hallway, looking at their phones, I could not take my eyes off this boy and the scene around him: flowers and bees, a pond, the golden hills to the east. And when a patch of sun broke through, spotlighting his little red jacket, I saw something else: that everything in front of me was part of a whole.

Because that boy's body was linked to the garden, and the sky overhead, through thousands of continuous, microscopic interactions. And those connections with his environment, though invisible from where I stood, were more fundamental to his health than any of the dazzling medical technology in the building behind me.

I realized that if he caught a raindrop, the water he swallowed, lent by the clouds, would be absorbed into each of his cells. That it would cushion his brain with cerebrospinal fluid, and his joints with synovial fluid; it would form acid in his stomach; it would course through his body, as blood.

I knew that his blood carried oxygen, being made and released by the garden's shrubs and trees, and drawn through his lungs from the air. That when he went home, the sandwich of wheat, peanuts, and grapes his mother gave him would contain carbon, pulled from the atmosphere as carbon dioxide. That those plants had converted the carbon into protein, carbohydrate, and fat that his body could use to grow, so that when he came into the clinic, we would mark his height and weight climbing higher, each time.

And I saw that those same atoms and molecules of water and oxygen and nutrients, now cycling through him, might have traveled through a tree, or a whale, or an insect, or a bird. They might have protected and nourished a baby in her mother's womb, or passed through an elderly woman on the other side of the world, living a life he couldn't imagine. Or through someone long gone, from a millennium ago.

Which meant that these elements were only borrowed from the atmosphere, not owned. That, as he played, the breath he exhaled and the moisture evaporating from his skin, mouth, and lungs were rising back into the clouds to be carried away by the wind. That the water in the urine he voided, that night before being tucked into bed, would find its way, eventually, back to the sea.

Because the water and air and heat that swirl around the globe, in clouds and rivers and oceans and wind, also flow through us, and every living thing. The atmosphere outside the windows of the clinic and hospital—the backdrop to my patients' visits—is *part* of their bodies, physically linking them to this planet and all its life. Children are creatures of this world. What we do to the Earth, we do to them.

But children are attached to their world by more than water and air. When my young patient finally caught a raindrop, he threw his fists in the air and ran to his mother, who was sitting on a nearby bench. She smiled and hugged him—and he went limp with mock exhaustion, knowing she would not let him fall.

The baby then squealed at being left out, and waddled toward them like a tiny drunk. The older boy turned, crouched down, and tenderly hugged his brother, who had collapsed into his arms. *It's okay*, I could see him saying, *it's okay*.

For a moment they were in a line: the mother's hands on her son's shoulders; him embracing his brother. Who could doubt, looking at them, that love passes through generations as surely as genes, linking us to everyone who came before, and everyone who will follow. Who could doubt what science now proves—and what I tell an audience of medical students every year, when I lecture about early childhood: that the love we show our children, like the water they drink and the air they breathe, becomes part of their bodies, literally shaping their brains, and hearts, and futures.

That same love, I hope, will save us.

Because it is the fiercest love that human beings feel. And no healthy parent, if they see their child is in danger, does nothing.

I once saw a mother, as she heard that her baby's odd and stubborn diaper rash was actually a rare life-threatening disease, *Langerhans cell histiocytosis*, collapse straight down as if the words had liquefied her bones. The cry that rose in that hospital room came from her deepest marrow, and etched into everyone it touched. I reached for her, too late; sometimes still I can feel her sleeve, brushing past my fingertips as she fell.

I feared, in that moment, she might die herself, dissolved on the floor between her baby's crib and the beeping machines of intensive care. But that mother, with much love and support, gradually looked up; she was able to listen, in ever-longer bits, as we described her baby's disease, how it was affecting her liver, and what needed to be done. Later she began to read about it; eventually she was calling specialists herself. After a few weeks she was an expert on what threatened her child and how to fight it. And that knowledge helped her stand again.

And every doctor, nurse, and technician who came into that room knew we needed her to stand; we could not save her baby with our tools alone. Her touch and voice were as vital to our patient as chemotherapy; her devoted advocacy for her daughter was essential to the medical care we gave. Defeating her child's illness required both science and love; neither was enough without the other.

I have seen this story repeat itself more times than I can count, with dozens of different diagnoses: diabetes, head injuries, autism, leukemia, congenital heart defects, all sorts of tumors and genetic disorders. Even in the best of circumstances, good parents often feel overwhelmed and afraid. But regardless of income, education, or personality, they never completely surrender—even when the odds seem impossible.

And that is why I wrote this book.

Because a disease of our own making is unraveling the fabric in which our children live; because our warming of the planet is already harming their physical and mental health.

Because we can still save our children from climate change's worst impacts—but we have less than a decade to change course.

Because even in the fastest-warming city in the United States, most of the parents I work with don't realize the urgency of this threat, and can't fight what they don't understand. Because those who do grasp what is happening sometimes feel too defeated and frightened to stand up, or don't know what to do.

Science is the foundation of decisions I make every day; it has saved millions of children over the past century from illnesses that were once invariably fatal. The little girl whose mother collapsed is now in grade school—because the doctors who cared for her believed, and responded to, the science.

But the science of what is happening to our planet, and what that means for our children, now demands our love. It demands that we look into the eyes of a teenage girl, grieving her future; at the face of a newborn, breathing smoke; at the blank expression of a child whose home, and town, are gone.

It demands that we look at a little boy in the rain, and see that everything about him—joy, beauty, and love—is bound to a body, rooted to Earth. That he is a physical being, subject to physical laws, as inseparable from nature as thread in cloth. That when we dig carbon from the ground and transfer it to the sky, we warm the world and endanger his life. And that the only way to save him is to stop destroying the atmosphere in which he lives.

Because while we think of climate change as a global problem, it is in fact a very personal problem, multiplied many times. It is a threat to the people we love most: the children we hold in our arms and promise to keep safe.

So I try, here, to tell the stories of some of these children—from American cities hit by worsening air pollution, extreme temperatures, and intensifying natural disasters. I use their cases to explain the medical and psychological problems these climate shifts are causing, and why children are especially vulnerable.

But as I think about the suffering of these families, I also remember that the climate crisis was a choice. James Black's granddaughter recently observed that had his warning been heeded forty years ago—had the fossil fuel industry led the development and expansion of nonpolluting energy sources like solar, wind, and geothermal—we might already live in a society powered by these technologies. That is not the path they chose.

These men were not comic book villains; they were real people who saw an immediate reward to themselves, but not the faces of children who would be hurt later on. I have stood in pediatric hospital wings named after them, funded by their money. Surely these men have children; surely they love their children like anyone else. Surely they can see the madness of this road.

A few months after my conversation with my teenage patient, another sixteen-year-old girl, thousands of miles away, rose to speak at a conference of the world's elite. Greta Thunberg called out those who suggest that the climate crisis is too big to solve because everyone on the planet contributes. "If everyone is guilty, then no one is to blame. And someone is to blame," the young Swedish climate activist said. "Some people, some companies, some decision-makers in particular, have known exactly what priceless values they have been sacrificing to continue making unimaginable amounts of money. And I think many of you here today belong to that group of people."

Later that year, tens of thousands of children around the world would rise with her in protest, marching through the streets and filling government halls. Shortly before the COVID-19 pandemic quieted their momentum, I ran into my patient at a youth climate rally downtown. She was holding a sign: *Save my future.* "It's a start," she said, and smiled. Her mother was right beside her.

Children are the least powerful people in society. They can't vote, and have no money; the youngest can't speak for themselves. But they are telling us something we need to hear. They need their parents to join them, and to fight with the same fierce protectiveness we would feel if they were ill, or injured, or being bullied. We can't put them on a rocket ship to another planet. We must save this unique and beautiful planet that is their home.

My family eventually left our neighborhood by the marsh, but not because of flooding. We returned from a movie one night and found neighbors gathered on the sidewalk; a fire truck, lights flashing, was parked in our driveway. Smoke was billowing from the back of the house, and firemen were running through it, the ground wet from their hoses. Our next-door neighbor, after a fight with her husband (and probably suffering from postpartum depression), had tied a bag of her family's laundry to the wooden fence that divided our properties, soaked it with gasoline, and thrown a match. The fence connected to my family's garage. Another neighbor had been walking by and spotted the smoke, saving our home.

Afterward I was very preoccupied with what might have happened if we had been in the house, sleeping, when our neighbor tried to burn it down. I asked my father about it. No one is going to burn down our house, he said; we are going to be fine. But what if somebody did try? What if it did catch fire? I insisted. My father looked at me. Then I would save you, he said. I would never leave you in a burning house. We moved to a different neighborhood shortly afterward.

Earth and humanity are in peril, and Superman isn't coming. The only heroes our children have are us.

Chapter One

BREATHING

"Isn't it funny," she says, "how you never think about breathing, until you can't."

I've pulled one earpiece out with a hooked finger so I can hear what she is saying. My stethoscope is on her daughter's chest; the round diaphragm rests perfectly over the left ear of Minnie Mouse, dancing on the baby's T-shirt.

"Sometimes we take things for granted we shouldn't," I say. Mom is standing by the exam table on the other side of the baby, one hand resting on her daughter's thigh. I put the earpiece back. A coarse sandpapery sound passes through the infant's lungs, a whisper of wheeze at the end. The baby contracts her belly just a little to push the last of the air out. I am looking into the ten-month-old's huge brown eyes, twin Jupiters studying my face. She sits as calm as Buddha while I follow her breaths: in, out. I smile. She is unimpressed.

"You're right, it's starting to affect her," I say, looping the stethoscope around my neck. "I don't know how much longer this will last,"—I nod to the window—"but it might be a good idea to get out of town."

Outside, ash falls like snow. The sky is black and gray. Everything I can see from our third-floor clinic is as dusted and lifeless as the moon; the air is churning with smoke and blowing debris. Stepping outside is like finding yourself on the wrong side of a campfire, but there's no getting up and moving. It's everywhere.

"It looks like the apocalypse, doesn't it," she says. The infant's car seat is at my feet. Mom had draped a damp baby blanket over it as a makeshift air filter for the trip to my office; it is speckled sickly gray. As she set down the seat an outline of soot powdered the linoleum floor, and I am now standing in it. *Avoid spreading ash*, the government warns us. *Do not allow children to touch or play in it.* She has picked up the baby from the exam table and they are watching flurries of cinder and haze whirl around the building like ghosts. I look at the two of them at the window and for a moment can't remember the world outside this dense, choking cloud; it seems like all there has ever been.

It is late August 2013. Two weeks earlier, a careless hunter in the drought-parched Sierra Nevada started what would become one of the largest wildfires in California history, the Yosemite Rim Fire, named for the nearby national park. Day after day, a relentless conveyor belt of wind carries the fire's thick, particle-heavy smoke directly to the Reno-Sparks area of northern Nevada, 150 miles northeast, home to nearly 400,000 people. My pediatric office, next to the region's largest hospital, is overrun with wheezing, coughing children.

Baby Anna is one of them. She has already been hospitalized twice in her short life with wheezing and respiratory distress. Her mother had asthma as a child, and Anna seems to have it, too. Even for those with normal lungs, the smoke has made the area virtually uninhabitable, with air pollution levels far above normal. Anna's mother is worried about getting more time off work, and hoping each day will be better. She's giving Anna breathing treatments and locking her inside their home, with duct tape and towels sealing windows and doors. But it is not enough to protect a baby with reactive lungs from air this bad. We are trapped in a smoke prison, and don't know our sentence. We wait.

Being a pediatrician is usually a wonderful job, filled with meaning and love. I look into the eyes of a lot of babies like Anna; I interact with children and their parents every day. We talk about their illnesses, injuries, and complaints, but also their drawings, their families, their dogs, their friends. I've known most of them since birth—in fact, one of the best parts of my job is how each day begins: meeting newborn infants who arrived overnight at the hospital. New life, and the joy a new baby brings to a family, always fills me with hope.

But I find it harder, lately, to delight in my young patients without a growing sense of guilt and worry. I know what these babies and children will face in their lifetimes. The Earth is warming, fast. The dangerous consequences of that fact are multiplying every year—making me fear for the generation I welcome to the world each day, and angry more is not being done.

Since Anna's visit to my clinic that afternoon, thousands of other wildfires have raged through California, just a few miles to our west. They have grown bigger and more explosive, devouring not just forests, but towns. Every summer and fall now, waves of smoke pass through my city, and more of my young patients cough and wheeze. In 2018, the Mendocino Complex wildfire would become the largest California had ever seen, darkening our skies for weeks. Only two years later, in 2020, the August Complex fire would shatter that record, becoming the first to burn more than a million acres. And in 2021 we spent not just days breathing smoke, as we did in 2013, but months, as both the Dixie and Caldor fires raged a few miles away.

When I look back today, I see that the Rim Fire was not an isolated event, as it seemed to us then; it was the beginning of a trend. It was a sample of the world we are creating for our children.

Donner Pass

I have not always been a pediatrician. When my own children were born I was an environmental planner, analyzing how rivers are affected by land use changes in their watersheds. I led a team of scientists and engineers collecting data on terrain, soils, and streams. We entered these data into computers to see what would happen to water flow if anything changed upstream—for example, if a strip mall replaced a forest. (This is what a "computer model" of the environment is—not a physical, miniature representation of something, like an airplane model, but a mathematical calculation of future reality.) We then made recommendations to local governments about what developments they should allow, or deny, to prevent flooding, preserve habitat, and so on.

I had an undergraduate degree in environmental studies and had nearly finished a graduate degree in forestry. But when my children were born I saw the whole world in them. My path changed. My husband had just finished graduate school and had a good job. I decided to stay home.

My own children's doctor sparked my interest in pediatrics. He would answer my questions about their ailments by pulling out medical textbooks and showing me his thinking. I loved the idea of a job that merged science with caring for children. So when crisis struck—when my world collapsed in a sudden and traumatic divorce and I found myself a single mom with three young kids—I had the beginnings of a plan. I looked at the roof I wanted to keep over our heads, took stock of my options and resources, asked my mother if she could help me with child care (which, thankfully, she did), and applied to the local university's medical school to become a pediatrician.

For years, I was completely exhausted, all the time. I fell asleep while reading to my children at bedtime or the minute I sat down on the couch after dinner. I often had to wake up hours before dawn to study or prepare for patient rounds. I asked classmates to record lectures for me so I could go to my kids' soccer games and teacher conferences. I made flashcards at the dining room table while helping the kids with their homework, sometimes setting my head down and instantly dozing.

When I graduated from medical school four years later, the closest pediatrics residency was at UC Davis Medical Center in California, but I couldn't move because I shared child custody with my ex-husband. So over the next three years I drove almost 90,000 miles between Reno and Sacramento to finish my residency and maintain half custody of the kids. I made hundreds of trips over Interstate 80's Donner Pass—one of the most treacherous winter roads in the country, and the infamous site where stranded pioneer families resorted to cannibalism to survive the terrible snows of 1846–1847. This beautiful, haunted mountain, which blocked the Donner Party's path and sealed them in horror, seemed both a literal and symbolic barrier between my old life and my new—and a stark warning about hubris, nature, and the ways children suffer for their parents' mistakes.

The members of the Donner Party underestimated nature and overestimated themselves. They reached the pass too late. They should have crossed before the snows, but were delayed because they took the awful advice of someone who cared more about his profit than their lives. More bad decisions and bad luck compounded this initial mistake. Almost half of the eighty-seven people who had set out from Illinois and Missouri the previous spring would die in the snow on the steep eastern side of the pass, frozen and starved, unable to cross the Sierras to the better life they had dreamed of in California.

Nature has its laws, and no empathy for us. It is important to trust the right people. Unimaginable tragedy is possible; in fact, it happens every day (I saw it every day, in the hospital). These were my fixations, crossing Donner Pass. I thought, somehow, they might protect me.

I was returning to Reno one night, inching my way down the eastern side of the pass in icy conditions, when I suddenly realized I had no control of my car—it was rotating. In less than a minute the rear end of the car was headed downhill, and I was facing oncoming traffic, sliding backward in the dark, toward what I knew from countless trips on this road was the cliff above Donner Lake, site of the pioneers' encampment. I pumped the brakes, but they didn't even slow me down; gravity pulled me backward, faster and faster, on the steep ice. Just as I was sure I was about to go over the edge, a soft thud and jolt told me I had hit something. The car stopped.

My breath fogged. A full moon lit the snowy peaks over my head and sparkled on the inky lake far beneath me, where a few cabin lights dotted the shore. I watched headlights coming toward me, slowly, from higher up the pass, mirrored in the icy road. Everything else was a sea of blackness, seeping through the windows, engulfing me. I sat paralyzed, wondering if whatever had stopped me would hold, suspended between here and not, watching the clouds of my breath in the frozen, lonely car. I imagined the dead families below, my morbid company on these dark trips, whose names I knew by heart; still reeling from the breakup of my own family, apathetic with exhaustion, I thought joining them was at least a poetically fitting end.

I had driven for hours in the storm to get this far, trying to get home to my children. I had nothing left. But in that moment, hanging on the precipice at Donner Pass, I could see life, exhaled; the simple beauty of just breathing became startlingly, crystal clear. I had been on call the night before in the pediatric intensive care unit, and had intubated a three-year-old with a severe lung infection. I knew so much about breathing. I knew so little.

I had realized, in medical school years before, that air and blood flow like rivers through the lungs, in tributaries of airways, veins, and arteries, nature mimicking itself as Earth's favorite forms are reflected in the structures of our body. I had seen that we are not separate or different from the environment I had studied in my prior career; we are entirely of it.

I had watched so many babies enter the world with that dramatic first cry, the first act of life, a breath. Now I saw that the last act, too, is a breath, and that breath and every one preceding it integrates us with, and binds us to, this world. We pull the atmosphere into our lungs, we breathe it out, we live.

The cars from uphill were getting closer. I had miles to go. Carefully, I turned the steering wheel hard right, touched my foot to the accelerator, and headed slowly downhill, toward home. Days later, driving past the same spot, I saw that only the bent end of a beat-up guardrail had stopped me from joining the ghosts below. I pulled over, looked at my car's paint scraped on the gray metal rail, and peered down at the lake. So often just a moment, or an inch, or a single choice separates us from tragedy.

Sometimes, though, tragedy and fortune are separated by a million moments, a million choices, made by a million people, steering whole countries and maybe the whole world in inspiring or terrifying directions, toward breathing, or self-destruction.

Mind the Atmosphere

In medical school it is easy to focus on the minutiae of the human body and neglect a bigger point: that we are here at all, breathing, is incredible. Every life a doctor tries to save, every person we have ever known and loved—in fact, every astonishing form of life on this planet—exists only because of the miracle of a thin band of gases blanketing the planet surface. We breathe these gases in and out, every day, without a thought. But just as a hair's difference can separate life from death, the presence of Earth's atmosphere is a result of astoundingly good—perhaps unique—fortune. We have not yet found another planet like ours, nor any other life, in the vast, lonely expanse of the universe.

The atmosphere is only as thick, relative to Earth, as a sheet covering a basketball, but it provides the air we breathe and the water we drink, shields us from the sun's heat and harmful radiation, and moderates temperature and precipitation. Combined with our perfect distance from the sun, the atmosphere creates a nurturing, habitable environment—the hospitable world we know.

Most of us are unconscious of this. We look at the atmosphere every day of our lives—on a starry night in the mountains, sharing a beautiful sunset on the beach, or watching a towering thunderstorm from our bedroom window—but we never see it. We breathe it into our bodies—while running for a soccer ball, bursting from a lake surface, or snoring softly in sleep—but we never feel it. Everything we have ever been, every moment and memory, and every hope we have for the future depends on the imperceptible ocean of gas in which we live.

While the atmosphere goes unnoticed, its destruction is insidious. The poisons we dump into it are invisible gases created by mundane acts—turning on the air conditioner, driving to an appointment, scraping uneaten dinner into the trash. Their effect is slowly cumulative over many years and huge populations of people; the harm of any single action is indiscernible, and the actual source of the emissions—power plants, gas wells, farms, garbage dumps—may be miles out of view. Seeing the damage being done requires knowledge, imagination, and a certain amount of trust in experts predicting something not immediately obvious. It's no wonder fossil fuel companies found it easy to confuse the public with propaganda denying climate change; the wonder is how their own hearts allowed it.

Our culture also interferes with awareness of hard physical truths. Distracted by the buzz and shine of modern life, we are disconnected from the natural systems on which we depend. Food comes wrapped in plastic from the store; water comes from the tap. The unseen web of air, water, temperature, and living things—the plants and animals who share this world and sustain us—rarely crosses our minds. We take for granted what we shouldn't.

Doctors share in this disconnectedness. We call the air we breathe "room air," as if our buildings create it. We understand well the mechanics of breathing: that only 21 percent of air is the life-giving oxygen we need; that the remainder is mostly nitrogen our bodies don't use, plus trace other gases. We know that when the diaphragm contracts, it creates a vacuum in our chest that pulls air through our mouth and nose, through the trachea and bronchus into ever-smaller branches, ultimately filling millions of tiny balloon-like sacs, the alveoli, clustered like grapes at the end of the tiniest airways. Capillaries surround the alveoli like fishing nets; these tiny blood vessels absorb oxygen from the alveoli and then send it to the heart to be pumped to organs and tissue. They also transfer carbon dioxide, a waste product of metabolism, back from the body into the alveoli so it can be exhaled.

We know so much about breathing. We know so little. The health of my patients, our children, our world, is changing—and all of us, now, can feel it. Isn't it funny how you never think about breathing, until you can't.

The Best of Times, the Worst of Times

This may seem an odd point in history for a pediatrician to be ringing alarms: in many ways, the health of America's children is the best it has ever been. In the mid-1800s, when the Donner Party set out across the plains, two of every five children died before their fifth birthday. In 1853 more than half of all deaths in New York City were children five and under. Losing a child was a tragically common experience, and fear of childhood illnesses hung over every parent.

It is tempting to imagine that parents' familiarity with grief weakened its blow. But a child's death has always been, and always will be, life's most wrenching sorrow. When Abraham Lincoln's third son, Willie, died in 1862 of typhoid fever, the depths of his wife's anguish made him fear for her sanity. Lincoln himself was unable to work for weeks. "My poor boy," he said. "He was too good for this Earth . . . we loved him so. It is hard, hard to have him die!"

Willie's physician, Dr. Robert Stone (who three years later would attend to the president after he was shot at Ford's Theatre), did everything he knew—administering concoctions of bark, mercury, and purgatives that, we realize now, were undoubtedly harmful. Willie could have been cured with a shot of penicillin, had it existed.

But I know that only because I read it in medical school. I have never seen a case of typhoid fever.

Or cholera, smallpox, polio, or diphtheria. Many of the infectious diseases that once killed children in large numbers, and which Dr. Stone would have known well, are now eradicated in the US, or nearly so. Through vaccines, water treatment plants, antibiotics, and modern prenatal and pediatric care, we have dropped our infant mortality rate—the percentage of babies dying before their first birthday—from nearly 1 in 5 in the mid-nineteenth century to less than 1 in 100 today.

But as we conquered these many threats to children's lives, a bigger, existential menace was emerging. Carbon pollution has been accumulating in the atmosphere since humans first burned coal for steam power at the start of the industrial age. It has forced Earth's average surface temperature gradually upward since 1850—with a sharp, alarming uptick in recent decades.

We see now that modern life is built on slow, ironic poison. Today's doctors can save a baby with a malformed heart, but driving to the hospital speeds up, ever so minutely, the end of his world.

I can do things Dr. Stone could not have dreamed of—I can see a child's kidneys with an ultrasound, or her brain with an MRI; I can know from a drop of blood whether a newborn has any of a host of diseases. But my profession faces potential horrors, down the road, that would shock even a physician of the Civil War.

How do I know? Because computer models are painting a picture of the world we face. They tell us that even if current international climate pledges are kept, the planet will warm 2.5 to 2.9 degrees Celsius over pre-industrial-age temperatures by 2100. Thanks to the falling price of renewable energy, that's an improvement from just a few years ago, when our "business as usual" path was predicted to raise temperatures by up to 4.5 degrees. Yet already, at 1.2 degrees of warming, the health impacts of climate change are wide-reaching and serious; the United Nations has said that current trends will create "endless suffering." Without more aggressive action, we may soon find ourselves envying the parents and doctors of the nineteenth century.

But why is a warmer world dangerous? Humans routinely experience temperatures from subfreezing to blistering. Why is a few degrees' rise in Earth's average temperature such a health threat, and why is it a particular threat to children?

Some of the health hazards of global warming are easy to grasp: we understand, as the pioneers did when they crossed the Great Basin Desert, that extreme heat can be deadly; we know that drought can destroy the crops that feed us. Others, such as the impact on breathing, are more subtle and complex. Climate change is increasing air pollution and allergens that trigger respiratory problems in many people. For these and other reasons described in the chapters ahead, global warming endangers nearly two centuries of public health progress and—if not aggressively and quickly addressed—is a threat to billions of people. Children are already suffering in this changing world—and are more at risk than their parents.

Children's Lungs Are Different

When Anna came to my clinic during the Rim Fire, she had been inhaling harmful, irritating particles and gases produced in huge quantities by the burning forest. The effect on her was greater than the effect on her mother or me. "Children are not just small adults," pediatricians often say. Their physiology and metabolism are markedly different, leaving them much more at risk from wildfire smoke and other air pollution.

Unlike our lungs, Anna's were constantly changing. Before she was born, her airways—the tubes that carry oxygen into her lungs every time she inhales—divided and extended like a growing tree, spreading ever outward into smaller and smaller V-shaped branches. At the ends of each branch, the alveoli—those tiny sacs where blood and atmosphere meet—began to form and multiply in much the same way.

But this process was far from finished at birth. As a newborn, she had only 20 million to 50 million alveoli, a fraction of the 300 million to 500 million she'll have as an adult; most of her eventual lung tissue had yet to be formed. By the end of adolescence, the surface area of her lungs should increase manyfold, ultimately reaching the size of a racquetball court.

The amount of lung development that occurs after birth is one reason children are so vulnerable to environmental harm, and underscores how inseparable a child's body is from the surrounding world. Every time Anna took a breath, the quality of the air she inhaled was literally shaping her lungs. In clean air, her lungs could undergo their amazing, continuous expansion. But when the sky is thick with smoke, as it was the day she came to my clinic, lung growth can be stunted. Air pollutants damage lung tissue, and inhibit the dividing and multiplying of alveoli. They also appear to "turn off" genes that are important in the development of the respiratory tract. As a result, children exposed to chronic air pollution tend to have smaller lungs that move less air.

If children are removed from high-pollution areas early enough in life, their lung function can gradually improve. But finding clean air can be a challenge: 36 percent of the US population now lives in counties with unhealthy levels of particulates and ozone—two pollutants linked to reduced lung function in children, and to global warming.

Lungs are one of the gateways through which bacteria, viruses, and fungi enter the body, so they have special immune cells to protect us from inhaled threats. But air pollutants, including those found in wildfire smoke, interfere with the development of lung immunity, leaving kids more susceptible to lung infections like bronchitis and pneumonia. For asthmatic children like Anna, who already have an increased risk of pneumonia, this is especially dangerous.

These impacts don't just magically resolve when children turn eighteen. What happens in early childhood often affects our health for life. Children with impaired lung function tend to struggle with growth, learning, and behavior. When they become adults, they have more respiratory and cardiovascular disease, raising the chance of early death. Injuries to young lungs can ripple through a lifetime, with a wave of costs not just to the individual, but to our health care system, economy, and society.

Some children may be harmed even before taking their first breath. Pollutants inhaled by pregnant women can cross the placenta and disrupt the development of a baby's immune system, lungs, heart, or brain. They can also stunt fetal growth and are a significant cause of premature birth (before 37 weeks' gestation), a complication that affects about 1 in 10 infants in the US. Those premature babies—especially if born at 32 weeks or sooner—then have underdeveloped lungs even more susceptible to injury from pollutants. They are much more likely to develop asthma and other lung diseases, and are more easily sickened by bad air.

But developing, vulnerable lungs are not the only reason children are more at risk. They breathe faster. Even at rest, Anna was breathing up to twice as fast as her parents. Also, though her lung surface area was smaller, it was larger relative to her size: for each kilogram of body weight, she breathed 50 percent more air than her mother or father.

More breaths, more lung surface area, and more air exchanged mean children are exposed to higher doses of air pollution, pound for pound, than are adults.

Children play more outdoors, and breathe air that is closer to the ground, where pollutants from automobile exhaust are concentrated. School-age children tend to mouth breathe more than adults, and younger noses don't filter pollutants as well. Both these traits increase their risk of inhaling particles—specks of solids and liquids suspended in air. And the cells that line their airways are easily damaged by these dangerous pollutants, making their lungs more vulnerable to inhaled threats.

In sum, children exposed to chronic air pollution are more likely to have smaller, stiffer lungs, be prone to asthma, pneumonia, and bronchitis, and die younger than people raised in healthy air. Yet many parents and pediatricians don't think of fossil fuels or climate change when caring for a wheezing, coughing child; the connection between these symptoms and the nearby coal-fired electric plant, or heavy traffic on the child's street, or our warming planet, is often out of sight, out of mind. When air pollution is visible and choking an entire city, though, it's a different story.

World on Fire

"What is happening?" a young father asked me in the summer of 2018, as he sat on the hospital bed of his three-year-old son. He was looking out the window at a blanket of gray smoke covering Reno for a third week; in front of him, gauzy shadows of downtown buildings came and went in the slow-moving haze.

The smoke had flowed over the mountains from the Carr and Mendocino Complex fires in Northern California, filling our pediatric ward with coughing children. His son and the other child in the room, my four-year-old patient Liam, were both on oxygen, delivered through a nasal cannula. The thin clear tube crossed from ear to ear, with prongs in both nostrils and a circle of tape on each cheek, holding it in place. Neither boy had ever wheezed, or needed oxygen, before.

I was standing between their beds, unsure if the father's question was really directed at me. Suddenly to my right, Liam rubbed his face, crying, and pulled the nasal cannula out. A *beep beep beep* signaled that his oxygen level had dropped; his mother calmed him and put the prongs back in his nose.

"You'll be home soon, buddy." I smiled, then caught myself: *soon* meant something very different to a preschooler than it did to me. He looked defeated as his small rebellion failed.

It had been five years, almost exactly, since Anna's visit to my clinic during the Rim Fire. Though smoke had pulsed through the city every fire season, this was our worst air quality crisis since.

For most of that month, we had not seen the sky; our world had shrunk to a few feet. All of our nerves were fraying. Driving to the hospital that morning I had been startled to see a teenage girl in a fast-food uniform emerge from the smoke as I passed. She was walking on the sidewalk, clutching a large white towel over her face. In the rearview mirror I watched her dissolve again as if she'd been an illusion.

Now, in the hospital room, I wondered what lay ahead of her and these two little boys. I looked at Liam, who normally couldn't stay still, limp in his bed, connected to the oxygen pump and monitor, a few spikes of sweaty hair stuck to his forehead. His mother was side-saddled next to him, gently rubbing his chest. He had snuck outside to play one day, and it had landed him here.

His roommate, the three-year-old, in his tiny pediatric gown, was staring at the television on the opposite wall. It was perched among glossy, bright-colored geometric shapes, beneath overhead light panels that looked like blue sky and clouds. From the TV a reporter in a windbreaker looked back at the little boy. She was holding a microphone in one hand and pointing, with the other, to a hillside of burning trees.

His father was still mesmerized by the *Twilight Zone* episode outside.

A moment ago, he had told me he was a landscape worker—and that this past month, with its ugly air and record-setting temperatures, had been hell. Then he'd coughed into a muscular arm bent over his face, both darkened by the sun.

What is happening. His question hung in the room. I glanced at the television, wanting the reporter to answer. But she wouldn't name this crisis, even as she pointed to it.

My head ached and my eyes stung after the short walk through the smoke from my car. With my background in forestry and environmental science, I was better equipped than most pediatricians to explain what we were living through. Since the Rim Fire, more than a hundred scientific studies have all pointed at the same cause. I knew why 129 million trees had died to our west, and now dotted the mountains like brown pox. Why heat was rising, making those dead trees explosively dry. How forest managers and utility companies had neglected this growing danger, allowing tinder to accumulate. Why wildfires were much bigger, longer lasting, and erupting more months of the year. Why his son and Liam were here.

I don't usually dodge difficult topics, even when parents and I disagree. But decades of propaganda and partisan warfare had made climate science political. Saying out loud what we could all see happening—out the window, on television, in this pediatric ward—risked something I treasured: my relationship with patients' families.

Yet as I stood there I realized another reason the words were stuck in my throat. Naming what gripped our city would make it real—and there was no medicine I could offer to fight it. For every other problem I discussed with parents, I could give advice or a prescription, an order for a lab test or a referral to a specialist. I could tell them what to do to help their own child. But for this, what could I say?

There was no way to protect their own children without fighting for all children. The only medicine we had was each other. The only thing we could do was speak up, together, against the corporations whose decisions had led us here. Because this smoke, these sick children, our learned helplessness and silence—all were a measure of their strategy, and the billions they had spent to get their way. And we were behaving like passengers in their car, failing to grab the wheel as they drove the world's children, faster and faster, down this slope.

"It's climate change," I said.

The father froze. Then he turned to his son and, smiling faintly, squeezed the small foot poking up through the sheets. "Yeah," he said. He knew the answer to his own question. He didn't want to say it, either.

But Liam's mother leapt at the realization that we were all thinking the same thing. "Haven't you heard, Dr. Hendrickson?" she asked, extending one palm toward the window, and the other to her son and his machines. "We don't need to worry. It's not real!"

Where There's Smoke, There's Illness

I said my good-byes and left—feeling suddenly, weirdly, hopeful. Instead of an argument, I had found an ally. And with just a few words of sarcasm, she had pulled aside the bleak curtain of those weeks to expose the obvious: the climate crisis wasn't only tragic, it was absurd.

I walked down the gleaming hallway, past more rooms of coughing and crying and beeping machines. I knew that pediatricians across the West were witnessing similar scenes. When I was a child, the country had never seen a wildfire bigger than 100,000 acres; now these "megafires" cursed cities from Seattle to Los Angeles to Missoula, and over 7 million children breathed smoke for part of every year. "Fire season" had joined "flu season" on my mental calendar of cyclical illness, and I had begun to dread the warmest months—previously a time when my schedule slowed.

But what, exactly, was in the smoke that was making these children sick?

At one level, the answer to that question was obvious. When the television reporter pointed at those burning trees, or later stepped through ashen neighborhoods, she was showing us what we were breathing. Megafires vaporize nearly everything they touch, and their smoke is packed with all of it.

In most fires, the people downwind never get a full accounting of what they've inhaled. But during the Rim Fire, special aircraft had collected samples of our dark air for analysis. Almost four years later, we were told that the smoke held 3 times more particle pollution than previously thought, as well as elevated levels of about ninety other compounds—including formaldehyde, benzene, and hydrogen cyanide.

Of these, particles—tiny fragments and droplets from trees, houses, cars, and whatever else has burned—are usually considered a wildfire's most dangerous pollutant. Smoke is dense with sooty debris, but the ash and cinders we see are not as harmful as what we can't: enormous quantities of microscopic "particulate matter." The smallest particles regulated by law, PM2.5, are less than 2.5 micrometers in diameter—only about 1/30 the width of a human hair. Wildfire smoke also carries huge amounts of even smaller "ultrafine" particles, less than 0.1 micrometer wide.

Bits of what once was, particles can be carried for thousands of miles in the wind. Because of their minute size, they can also be pulled deep into the lungs; the smaller the particle, the more invasive and hazardous for human health. The finest particles easily cross the alveoli, enter the bloodstream, and wreak havoc on multiple organs, including the heart and brain.

Particle pollution is not exclusive to wildfires. Fossil fuels—whether burned in coal-fired power plants, diesel trucks, airplanes, or cars—are the most important source. Over the last half century, the Clean Air Act reduced emissions from tail pipes and smokestacks across the country. Yet scientists estimate that in 2018 an astounding 13 percent of all deaths in the United States, and 8.7 million deaths worldwide, were caused by particles from fossil fuels.

Now, as wildfires intensify, airborne particles are surging, especially in the western US, becoming a major health threat tied to the climate crisis. Between 2017 and 2021, both the severity of smoke pollution and the number of Americans exposed to it rose; at the end of that period nearly 64 million people lived in counties badly affected by wildfire particles, the most ever. Almost all the cities with the worst "short-term particle pollution" are in the West. But as wildfire smoke from Canada engulfed the Northeast and Great Lakes regions in 2023, giving New York, Chicago, and other previously spared cities their worst air quality in history, many Americans saw that the health impacts of burning forests won't stay confined to western states.

The implications of these trends are sobering. As many as 3,000 Californians over the age of sixty-five were likely killed by PM2.5 during the record wildfires of 2020—many times the number of people who died in the flames. In adults, particle pollution increases the risks of heart attacks, strokes, lung cancer, chronic obstructive pulmonary disease (COPD), diabetes, kidney disease, and dementia. COVID-19 appears to be more easily transmitted, and causes greater illness and mortality, in areas with higher particle pollution; Reno saw an 18 percent rise in cases tied to the 2020 fires. Exactly how much worse the pandemic's toll was in the West, because of smoke, is unknown. What is clear is that when particle levels jump in a city, the non-accidental death rate does, too.

Very few studies have examined the impact of wildfire smoke on children. It's a difficult question to answer fully because smoke's contents are complex and differ from one fire to the next. But three studies that focused on the particulate matter in smoke—from Southern California wildfires in 2003, 2007, and 2017—found that it alone had a significant impact on children's respiratory health. Children and teens downwind suffered higher short-term rates of asthma, bronchitis, pneumonia, sinus infection, and allergy symptoms. The bigger the dose of smoke a community got—the more days of bad air, or the higher the particle levels—the more respiratory illness among its kids. One of the studies found that children under five years old were most affected.

Based on these findings, I assumed that particles were the main culprit behind Anna's and Liam's coughing, wheezing, and shortness of breath during the fires. Further evidence would come a few months after Liam's hospitalization, when health researchers from our state university showed me local air quality data plotted against daily all-ages medical visits for asthma. Every major wildfire leapt from their graph: a wave of particle pollution, rising and falling over a string of days, and a second wave, of illness, mirroring the first but lagging by a day or two. I knew my young patients were somewhere in those curves.

When the smoke cleared, so did their symptoms. But I have reason to worry that their exposure to wildfire particles might have lasting effects. In recent research from Stanford University, blood tests on school-age kids in central California revealed that particulate matter from a single, nearby 2015 megafire had actually changed their DNA. Specifically, the pollution had altered a gene needed to form T cells, a vital part of the immune system. As a result, the children made fewer, less healthy immune cells, raising their later odds of allergies and infections.

Not all particles are created equal. Particles from wildfires cause more lung inflammation and hospital admissions than those produced by fossil fuels. In fact, a multiyear study in San Diego County found that PM2.5 from wildfires was roughly *10 times* more likely than particles from other sources to cause breathing problems in children, especially if they were under five years old. And particles from megafires seem to inflict more damage than those from smaller, controlled fires. This may be because megafires' intense heat and destructiveness produce more ultrafine and toxic particles, including "black carbon," an irritating microscopic soot that binds to chemicals and heavy metals in the smoke. Climate change is thus increasing not just the quantity of particles but also, apparently, their toxicity.

This creates another outsized threat to the young. Take, for example, polycyclic aromatic hydrocarbons (PAHs), cancer-causing chemicals that are often bound to particle pollution. Although PAHs are a threat to everyone, an adult's liver contains enzymes that break down these chemicals so they can be excreted. Other adult enzymes repair pollutant-damaged DNA in our cells, reducing the odds of cancer. But in fetuses, babies, and children, these defenses are immature, so exposure to the same "dose" of PAHs is more likely to cause harm.

Some of my patients' families have tried to escape smoke's dangers by moving from the West; since megafire pollution disperses across the country, however, and since wildfires are spreading to other regions, it is difficult to avoid completely. Others forget about the smoke when our skies return to blue, or believe next year will be better. The rest of us are trying to adapt to our new, seasonal reality. Even before COVID-19, I recommended N95 face masks for my patients during wildfires. If they are too young to wear one, I recommend a fabric cover over the car seat. Air purifiers are now found in any home that can afford them, though some families make their own from instructions available online. Schools keep kids inside on "smoke days" or cancel classes altogether; some have installed special ventilation systems to filter both viruses and particulates.

We are all doing our best to cope. Yet most parents in my clinic are understandably anxious about the fires. No one really knows how repeated exposure to smoke is affecting their children's long-term health. Will its toxic chemicals increase childhood cancers? Will its particulate matter cause local kids to have more heart disease as adults? Data on these questions are basically nonexistent. But we have good reason to believe that wildfire smoke causes more health problems for children than just brief coughing and wheezing. That's because of a large body of research on particles from other sources—namely, fossil fuel burning.

Inhaled Particles Cause Long-Term Harm

I noted earlier that a child who regularly breathes particle pollution will tend to have smaller, stiffer lungs. We know this largely because of the Children's Health Study, which in 1993 began following the effects of urban air pollution on fourth graders from a dozen Los Angeles–area communities. For eight years, researchers tracked the children's lung function against the levels of air pollutants they were exposed to in each town. By age eighteen, kids who were raised in high-PM2.5 areas were about *5 times more likely* to have "significant deficits" in lung capacity than those raised in low-PM2.5 areas.

Children who had moved to neighborhoods with cleaner air during the study recovered some of their lung capacity. But for young adults who had not been so lucky, lung growth and breathing were unlikely to improve after age eighteen. As a group, they would have more trouble exerting themselves, thus limiting their job options. Their impaired lungs also increased their odds of COPD, hypertension, heart attacks, strokes, and dying too soon. In other words, the particles in fossil fuel emissions—from cars, trucks, and factories—had silently and insidiously harmed these kids for life.

We have every reason to believe that wildfire particles would do the same, although exposure to them tends to be much more intense—with levels many times higher than what the LA kids experienced—and seasonal instead of daily. We don't know if "episodic but terrible" pollution is better or worse for lung development than "chronic but less severe." I always hope that the clean air Nevada still sees for much of the year allows my patients' lungs to heal, much like moving to a less-polluted town helped the children in Southern California.

Since that study was published in 2004, pediatricians have learned a lot more about particles and the damage they cause. We now know that not all age groups are equally affected: the younger the child, the greater the danger. It's clear that low-income and minority children are most heavily exposed to particles and the toxins they carry, such as PAHs, because their neighborhoods are where freeways, power plants, and factories are often sited. And we have growing evidence that the lungs are not the only part of a child's body that can be hurt.

Among my patients, newborns appear to be most at risk. Babies exposed to high particle levels are more likely to die in the first year of life, especially of respiratory causes. Particle pollution increases the incidence of sudden infant death syndrome (SIDS), bacterial pneumonia, and viral bronchiolitis, a lung infection that strikes babies hardest. Reducing particulate matter, conversely, improves infant survival. A Columbia University analysis found that converting America's electric power plants to renewable energy sources would lower particle-caused infant mortality about 10 percent over the next three decades—saving 2,500 babies' lives.

Even exposure *before* birth can cause harm. When particles are inhaled by a mother in early pregnancy, the odds of congenital heart malformations increase. Prenatal exposure has been tied to a higher risk of cancer before age six, childhood hypertension, and juvenile obesity, perhaps because of changes to DNA or to the endocrine glands that govern our metabolism.

Particles enter and inflame the placenta, which is probably why they raise rates of miscarriage, premature labor, and low birth weight. While a baby can be born too early or too small for many different reasons, air pollution is a major trigger. In 2019, *2 million* premature births around the world were attributed to PM2.5 exposure alone. Small, preterm babies then have more respiratory, cardiovascular, and developmental problems. In fact, complications of premature birth kill more infants and toddlers, globally, than any other cause.

Perhaps the most startling research of recent years, though, shows that children's minds, too, can be wounded. Those who live in high-particle areas, or whose mothers were pregnant in such areas, have a greater chance of a learning disability, attention problems, and the disorder most parents worry about, autism. One likely but disturbing reason: particles invade the brain. Mainly, they target and inflame the *prefrontal cortex*—the part of the brain right behind the forehead—where they migrate after being inhaled through the nose.

That's a shocking finding, one I wish weren't true. But when scientists scanned the brains of a group of children in Mexico City, notorious for its bad air, they found prefrontal inflammation in more than half of them. The children scored poorly on cognitive tests compared to children from a low-pollution city, whose brain scans were almost all normal. Later research confirmed that particles, encased in Alzheimer's-like lesions, were embedded in the prefrontal area of Mexico City's children and young adults.

Which has enormous implications. The prefrontal cortex is central to our personality and behavior, as well as our "executive function"—the ability to plan, focus, remember, and prioritize. The study's authors concluded that particle pollution had significantly impaired these young people's thinking. And it had sowed the seeds of later dementia, which has also been tied, in elderly people, to particles in the brain.

Though these tiny pollutants like to cluster in the prefrontal cortex, they can reach any part of the brain via the blood, which they enter from the lungs—either the child's or, in the case of a fetus, the mother's. Regardless of their path, when particles and the toxins that stick to them invade brain tissue, or when they cause general inflammation of the body, they can disrupt the normal development of a still-forming brain.

That's the theory, at least, as to why so much new research points to particle pollution as a risk factor for both autism and attention-deficit/hyperactivity disorder (ADHD). In a multiyear study of almost 300,000 children in Southern California, for example, prenatal exposure to PM2.5 was found to significantly increase autism rates. Boys, who are diagnosed with autism 4 times more often than girls, seemed more susceptible to the pollutant's effects. And a recent study of over 164,000 school-age children in China added to evidence that long-term exposure to fine particles boosts the odds of ADHD. Though autism and ADHD are complex disorders with multiple causes both genetic and environmental, a growing body of science indicates that air pollution—caused by fossil fuels and worsening due to climate change—is contributing to the rise of these conditions.

I find myself explaining this now, every time a parent asks me about the disproved link between vaccines and autism. We have no evidence for that, I say. But let me tell you about the particles inhaled from oil, gas, and coal. Let me tell you why wildfires and smoke are intensifying around us; how they are spawned by these fuels that warm the world.

At least, that is what I've said ever since that day in the hospital. Ever since Liam's mother pointed to the window, and to her son.

Good Ozone/Bad Ozone

I wouldn't see him again until the following spring. He was sitting on the exam table as I entered the clinic room, stripped to his Pokémon underwear and white socks, a halo of sunlight around him. "Doctor H!" he said.

"Liam!" I answered. Before I could close the door, he had climbed down to hug his mom, his sister, and me. With his chart under my left arm, I wrapped my right around his little body. When he darted back to the table, I gave him a boost. He was all knees and elbows now; his toddler pudge was gone.

He was there for his five-year checkup and wanted to tell me a story. "Wow," I said, not understanding most of it but getting the gist, glancing at his mom for translation. Liam struggles mightily with speech.

He struggles with other problems, too. He was born early, and small, with a congenital heart defect that required surgery. He has trouble with learning and paying attention. His mother once asked me if any of these issues were caused by the smoke she'd breathed during her pregnancy. She had been in her first trimester with him during the Rim Fire.

I had to say I didn't know. No lab test or X-ray can answer her question. We can only understand environmental harm by comparing large groups of people who were exposed to similar groups who weren't, and then looking at the differences between them. So while I could tell her that these problems were more likely in exposed children, I could never say that smoke was the specific cause of *his* problems.

But I wondered, too.

I looked in his ears with the otoscope, and asked if they had any plans for the upcoming summer. "Stay out of the hospital," his mom joked. Had she seen the air quality app that could alert her when kids shouldn't be outside? I asked.

"I have the old-fashioned app for that," she said, gesturing to her own eyes and nose.

She was right: it was hard to miss the sight and smell of smoke. But on that warm, smoggy day I was thinking of another pollutant increasing due to climate change—one that Liam had breathed throughout his life, and that has brought Anna back to my clinic more than once: ozone.

Ozone has a more complex role in human life than most pollutants. When high over our heads in the atmosphere's famous ozone layer, it is not a pollutant at all: it shields us from the sun's ultraviolet radiation, protecting the health of all living things. Without this "good ozone," Earth would be lifeless. But at ground level, where children breathe, ozone is a hazard. A key part of urban smog, ozone inflames the respiratory tract, and is a potent asthma trigger.

Where does the "bad ozone" in smog come from? The answer reveals an ironic trap of modern life. When Liam's mother drove him to his checkup, she was being a good parent. Yet her car burned gasoline, and emitted exhaust filled with toxic chemicals. Once those pollutants were released, they mingled in the hot city air, where some of them—nitrous oxides and volatile organic compounds—reacted with one another to form ozone.

Across the city, thousands of other cars, trucks, and planes were spewing the same "ozone precursors," which then met and recombined in the sunlight. Our use of fossil fuels ensured this; the oil companies had ensured this, with their decades-long fight against clean energy. To get to Liam's appointment, she had no choice but to pollute the air he breathes.

The chemical reaction that forms ozone is very sensitive to temperature: warm air speeds it up. So as the city's thermometer climbed that day, its ozone pollution did, too. This was not a new problem for Reno, or any urban area. Cities are where most vehicles, power plants, and other sources of pollution are found. And their buildings and streets create zones of warm, stagnant air known as "urban heat islands"—perfect conditions for ozone formation. It's no surprise, then, that ozone pollution has plagued cities for years.

What is new is the temperature itself. As the world warmed over the past decade, urban ozone levels climbed, reversing years of progress under the Clean Air Act. Of the twenty-five most ozone-polluted cities in America, seventeen got worse in the record-hot years of 2015 to 2017. These included most of the country's biggest metropolitan areas, such as Los Angeles, Chicago, and New York, but also small cities like Sheboygan, Wisconsin. Tens of millions of children like Liam were exposed to more high-ozone days. Experts laid the blame squarely on climate change.

Between 2019 and 2021, ozone levels improved in many areas due to decreased commuting during the COVID-19 pandemic, stronger emissions standards for cars and trucks, and the closing of some coal-fired power plants. It was a bit of good news that proved how much control we have over the problems caused by fossil fuels. Western states lagged, however, because of climate change, which is causing faster warming and more air stagnation in arid and high-elevation areas. In fact, Reno's ozone levels continued to worsen during this period; my young patients now live in one of the most ozone-polluted cities in the country.

Climate change is also increasing my patients' ozone exposure via wildfires, which emit huge quantities of ozone precursors. The Rim Fire generated air pollution equivalent to the annual emissions of 2.3 million cars. When those gases mixed in the fire's heat, ozone formed and was carried in the smoke downwind. It's estimated that over two thousand pediatric emergency room visits a year are due to such wildfire-produced ozone.

Ozone's danger lies in its chemistry: a single molecule consists of three atoms of oxygen, O_3, instead of the two-atom version, O_2, that our lives depend on. Yet what a difference that atom makes. Instead of keeping cells alive, ozone irritates them—specifically, the epithelial cells that line our respiratory tract.

Children's airways are especially sensitive. When Liam left my office and stepped outside on that smoggy day, he faced the risk that ozone would inflame his nose and lungs, causing them to make mucus. And that his lungs' irritated airways would spasm and constrict, resulting in wheezing, coughing, and a "breathing through a straw" feeling—the symptoms of asthma.

A child with asthma, like Anna, has more "reactive" airways. For her, even a small increase in ozone can cause distress. Liam hasn't had asthma in the past but, like anyone, could still have an asthma-like attack—as he did during the fires—if pollution is bad enough. Worse, chronic exposure to ozone could cause him to *develop* asthma. The evidence for this comes from studies showing that children who play outdoor sports in high-ozone areas are more likely to get the disease, and those living near major roads, where ozone is higher, have 50 percent greater odds of acquiring it.

Like the particulate matter Liam inhaled during the fires, ozone pollution may, over time, reduce his lung function, measured as the amount of air he can expire in one minute. It may also affect his learning: even a minor rise in ozone levels can trigger a huge jump in school absences for asthma and other respiratory illnesses. As was the case with particle pollution, higher ozone levels are associated with certain birth defects, lower birth weight, prematurity, and SIDS, and boost non-accidental mortality across age groups, mostly from lung and heart diseases.

Of course, if Liam's mother can someday get to my clinic *without* burning fossil fuels—if *everyone* stopped burning these fuels—ozone pollution would essentially disappear from the list of humanity's problems. And the exhaust from these fuels would no longer add to the warming of the planet.

Recent trends in the US give reason for hope. But if emissions aren't cut further—or if those gains are reversed by politicians friendly to the oil and gas industry—we know what comes next: the more the world warms, the worse ozone pollution will be. These higher levels will be largely invisible, but not benign. Emergency rooms and hospitals will fill with wheezing, coughing children; school desks will sit empty; and countless families will lose a loved one sooner than they should have—as I have seen, firsthand.

Medical Mysteries

I finished my pediatrics residency in Sacramento trailed by small ghosts: children I cared for in the hospital who did not make it, whose last moments I will always carry. One of them was a fourteen-year-old girl named Ruby who succumbed in the emergency room to a severe asthma attack. I can still see my hands clenched together over her sternum after CPR, the team of people who had tried to save her stepping back as the time was called. She had warm brown skin and hair pulled into a small ponytail; as I withdrew, I saw a child's sequin purse resting beneath the gurney. She had been playing jump rope with friends. I cannot forget the sounds of her parents' grief.

In her purse was an albuterol inhaler, which she had been using when paramedics arrived. It should have relaxed the muscles around her airways, allowing them to open. But if an asthma reaction is bad enough, as it was for her, it can make breathing impossible—the body literally suffocates itself with inflammation. Though fatal attacks are rare, roughly two hundred families a year lose a child this way. Poor and minority families are disproportionately affected; African American children are most likely to die from the disease.

It is one thing to cite those statistics. It is another to have it happen in your hands.

All that week, the ER had been humming with nebulizers—machines that turned albuterol into mist, each beside a wheezing child who was breathing the medication through a mask. I didn't understand the reason for all those sick kids at first. My shift went from dawn to dusk, and the ER had no windows. I hadn't seen what they were breathing outside.

And that wave of illness didn't strike me as too unusual when it began, because asthma was hardly rare. It was one of the top reasons kids were brought into the ER or admitted to the hospital—not just at UC Davis, but everywhere. The statistics had been pounded into me and every pediatric resident: asthma was the most common chronic pediatric illness in the country; over 6 million children had it. Kids were more likely than adults to get the disease, and suffered more frequently from asthma attacks.

But after Ruby died—after I stood beside the attending physician as he spoke to her shattered parents—I stepped out to get some air. The city was muggy. The ambulance bay felt like a concrete sauna, and my scrubs stuck to the sweat on my back. I looked down the busy street, filled with cars whose drivers didn't know what had just happened, and saw a layer of yellow-brown haze, looming darkly above the horizon like a thief.

Until the 1960s, some doctors had believed that the wheezing of asthma was a suppressed cry, a sign of a child's emotional distress. Not until the

1980s did we begin to understand that asthma is a chronic inflammatory disease of the lungs, and that air pollution makes it worse. Handheld inhalers to deliver asthma medication had been invented by that time, but were not widely available, and albuterol—now found in every pediatric and school nurse's clinic in the country—would not reach the market until 1981.

Yet asthma in those years was far less common. Though I see it nearly every day in my clinic, that was not so for pediatricians of the past. "Prior to 1960," one medical historian noted, "most textbooks of pediatrics did not regard asthma as common let alone epidemic." The percentage of American children with the disease almost *tripled* from 1980 to 2010—the period that framed young Ruby's life.

This rapid rise was part of a pattern. All *atopic diseases*—asthma, allergies, and eczema—jumped dramatically in those decades. These illnesses, characterized by an overreaction of the immune system to irritants, often occur together in the same person. Asthma and allergies are particularly intertwined. Most asthmatics have some type of allergy, and allergic reactions frequently trigger their asthma attacks.

That was on my mind when the paramedics rushed in from the ambulance bay with Ruby, shouting out her story as they unstrapped the gurney and lifted her over to us. Was she having an asthma attack, a severe allergic reaction, or both? Her mother was alongside, frantic. *Can you tell me what happened?* I asked. Her friends had come running into the house, she said, and told her Ruby couldn't breathe. *Does she have any allergies?* Yes, she has hay fever, and takes Claritin for watery, itchy nose and eyes.

Allergy symptoms, when a child has them, are triggered by *allergens*—proteins such as those found in pollen, mold, dust mites, or peanuts—that have been touched, eaten, or inhaled. Allergies are, like asthma, very common in children; these days, pediatricians' schedules are often filled with allergic complaints. The World Allergy Organization estimates that *nearly half* the world's children are now sensitized to at least one allergen, an almost 1-in-2 prevalence confirmed by studies of American kids.

Why asthma and allergy rates have skyrocketed over the past half century, though—why there are so many more children like Ruby, and Anna, the asthmatic baby who was sickened by the Rim Fire—is one of medicine's biggest mysteries.

The medical detectives trying to solve this puzzle have one big clue: these illnesses are not increasing equally everywhere. Atopic diseases are less common in developing countries and rural areas. But as countries develop, or as people move to urban areas, their prevalence grows. What is it about cities and industrialization that causes this?

Part of the explanation may lie, surprisingly, in the nineteenth century—in the era of the Donner Party—long before Ruby and Anna were born. Before clean water and public sanitation, infections raged, but atopic diseases were rare. Some research suggests that our immune system needs exposure to a rich variety of microbes to learn not to overreact to foreign proteins. Children in rural areas get such exposure, as our ancestors did, by spending more time in dirt and around animals. This *hygiene hypothesis* argues that soaring asthma and allergy rates are, in part, an unintended consequence of cleaner habits.

But many researchers see the hygiene hypothesis as an incomplete explanation for why asthma and allergy rates rise with industrialization. They point to the most obvious difference between cities and everywhere else: air pollution and allergens are worse in urban areas, and climate change is fueling them both.

It makes sense: development has always meant electricity, transportation, and buildings powered by coal, oil, and gas. Those fuels, running our cities and degrading the air, make asthma symptoms more frequent and more severe. Their emissions likely cause asthma to develop in genetically susceptible children and appear to have *epigenetic* effects: they can alter how a child's genes are expressed, increasing the odds of respiratory illness. Viewed through this lens, climbing asthma rates might be explained by the simple fact that, each year, more families were living in polluted air: over 80 percent of Americans were city dwellers in 2010, up from 74 percent in 1980.

Among children like Ruby—from a low-income family living in a heavy-traffic area, and thus chronically exposed to higher levels of ozone and other air pollutants—asthma rates are highest. Because of the poor-quality air she'd breathed all her life, she was more likely than a child from a wealthy neighborhood to have reactive airways, to be rushed to the emergency room on that bad-ozone day, and to die.

That difference, caused by indifference, is evidence. It suggests that air pollution played a role in asthma's rise, and that Ruby might not have had asthma at all, if not for Sacramento's smog—made worse because of rising temperatures.

Ruby's odds of hay fever were also increased by fossil fuel pollution and global warming. Cities and suburbs create ideal growing conditions for pollen-producing plants. These areas have high levels of carbon dioxide, which, in addition to being the most important greenhouse gas, stimulates plant growth. And they are not only warmer than rural areas—that urban heat island effect—but warming faster.

Consider ragweed, whose pollen is a key cause of hay fever and asthma attacks. Rising carbon dioxide levels and temperatures are boosting its growth, extending its range, lengthening its growing season, and increasing the amount of pollen produced by each plant. The pollen itself has grown more potent, containing more of the protein that causes an allergic reaction. Ragweed pollen levels in the US have *doubled* over the last century, creating legions of allergy sufferers.

No wonder that every year, like clockwork, the media run a story about another "record-setting allergy season." Yet few of the parents I know who buy allergy medications for their children, or struggle through the night with their child's allergy-triggered asthma, have any idea these symptoms are linked to fossil fuels and climate change.

And ragweed is only one example. The "poison" in poison ivy is becoming stronger. Allergic reactions to insect stings are increasing as warmer winters alter insects' life cycles and venom. Tree species such as oak and hickory—common allergy triggers—are replacing evergreen species that rarely cause allergies. Mold is proliferating in thousands of flooded homes.

Even the rise in food allergies—whose prevalence in children increased 18 percent between 1997 and 2007—may be related to climate change. Some scientists argue that the spike in peanut allergy is linked to that allergen becoming more potent. And certain pollens, including ragweed, can cause cross-sensitization to food allergens, a phenomenon dubbed *pollen-food allergy syndrome.* Studies indicate that up to 43 percent of children with pollen allergies had such food cross-reactions. As pollen levels surge, these related food allergies are climbing, too.

Ruby's hay fever likely played a role in her asthma attack. When pollutants and allergens mix, they can interact in ways that worsen the body's inflammation. Pollen bound to particulates, for example, can penetrate more deeply into the respiratory tract. Allergens may cause a more intense response in tissues already inflamed by ozone, and vice versa. These synergistic reactions are another reason allergies are worse in polluted urban areas, and can make what might otherwise be a manageable asthma attack into something much worse.

Maybe asthma and allergy trends are not a mystery so much as another example of our inability to see the obvious: fossil fuels poison the atmosphere, including the air we breathe. Our children pay a staggering price. We do a poor job of counting these children, and the full weight of what they bear. But they are not nameless, or faceless. We look into their eyes, every day.

Anna's Future

In May 1847 a young survivor of the Donner Party, Virginia Reed, wrote to her cousin back home in Illinois. She and her siblings had been rescued from their horrific camp in the Sierras only two months before, and were recovering in the safety of central California. Virginia would go on to have children of her own and live a long life, but the trauma of her family's ordeal was always a part of her. Her letter, written that first spring, reads like a dam breaking. "Oh Mary," the thirteen-year-old said, "you don't know what trouble is."

Virginia walked past what is now my home—then just a meadow along the Truckee River—more than a century and a half ago, on her way to the mountains. I can stand in the steps of this long-ago girl and look at the peaks that towered in front of her. I sometimes wonder what she felt. Snow was already falling when her family set out for the pass, but they continued toward it; almost half the Donner Party members were babies and children, but their parents did not appreciate the dangerous weather ahead. *Don't go,* I always think. They can't hear me, and the past is unchanged.

It is easy to look back now and question all the misjudgments and mistakes the Donner Party made. Though the world is a very different place than the one they knew, human beings are much the same.

The smoke coming from the burning forests is, like the snow the pioneers looked at but didn't see, a sign of things to come. I wonder: If a well-meaning stranger from the future could show us where our path leads, and tell us to change course, would we take her advice? If we could really visualize where we are headed with our children, would we act more urgently?

Because while we can't undo our own decades of denial and inaction, we can still shape what happens next. We have solutions that could save today's children from the worst. What we don't have, anymore, is time.

I drove over Donner Pass again not long ago. Descending its western slope, I came upon a traffic jam and slowed. A helicopter towing a bucket of fire repellant arced overhead, releasing its load on a burning ridge. Crews had already arrived to try to contain the blaze, which was raging through dead trees along the interstate. As we inched past, angry orange flames flicked at our cars like accusation, while torrents of gray smoke churned skyward, pouring east.

Through the car window I watched the smoke go, knowing that within a few hours it would cross the lee side of the mountains and settle over the people of Reno. I thought about the children of my town. I thought about air pollution everywhere, increasing because of the climate crisis, and all the damage—stunted lungs, malformed hearts, being born too early or too small, suffering with asthma, dying too young—that didn't need to be. I thought about the possible world.

Breathing is the most fundamental necessity of life, the first thing a doctor tries to protect and restore in a coding patient. It ties us to the atmosphere, to our planet, to existence. Fossil fuels and global warming are endangering the most basic of my patients' needs. Children can't live healthy lives without healthy air.

Yet breathing, vital as it is, is only the beginning of the story. Rising temperatures are themselves a threat to children's lives. And in Reno and most other cities in America, thermometers are climbing, fast.

After Anna's visit to my office that day, I stood in front of the exam room and watched her leave, waving good-bye to my small patient. She was facing backward in her car seat and being carried by her mom, who suddenly looked as if the world was asking too much. Those big brown eyes stayed fixed on me until her mother paused, halfway down the hall, and covered her car seat with the ash-stained blanket. Just before she disappeared, Anna raised a hand in good-bye. *You don't know what trouble is,* I warned myself, and they headed back into the smoke.

Chapter Two

SILENT AND INVISIBLE

"His case affected me more than any other call I've been on," he says, "and I've been doing this awhile."

He's speaking to me from a fire station in north Phoenix, not far from where it happened. It's been almost a year. "When the call came in that afternoon, I worried right away about what we'd find. It's one thing when adults risk their own lives," he says. "This was something different."

He had arrived at the trailhead parking lot that day and found a guy pointing to the mountain, screaming, *That way, a quarter mile!* A crew got there right behind him and they all took off running, up the trail. Chilled IV fluid and two bottles of water were in his pack, heavy and sloshing and slowing him down; one bottle in case the boy could still drink, one so he wouldn't need rescuing himself.

This is how he spends his summers: rescuing people from the heat. The department had almost 280 of these calls that year. "Some weeks it's like, man, *another* one?" Many are hikers, but not all. He's resuscitated people who passed out in their gardens, in their bedrooms, at their construction jobs; some, on the streets where they live and sleep. Just a month before we spoke, he had saved a baby locked in a car at midday. Her mother was pressed on the opposite side as he shattered the window, releasing a blast of hot air from inside. A paramedic in Phoenix knows all about margins, and error.

Churning up the slope that day, he remembers, felt like moving through wool—like one of those dreams where you're trying to run, but can't. His sunglasses slipped on his nose; he pushed them up, kept his eyes down. Beneath the pack, his shirt was soaked. He caught himself wanting, irrationally, to fit beneath a desert shrub—any shield from the sun. His partners, huffing alongside, were in the same shape.

They rounded a curve and saw the tree, just ahead, rising from a gully in the hillside. It was an old mesquite, with multiple trunks half-defiant, half-uncertain about where they found themselves—scraggly and bent, marking better years and worse. One trunk pointed sharply downhill, toward the arroyo below, as if debating its hold on this slope of rock and dust. The tree's leaves were tiny and cautious. Its taproot, he knew, went deep into the Earth.

Beneath the tree was a small half-circle of shade. A lizard darted from it as their footsteps approached, stopped mid-trail, and tilted its head toward the sun. Not enough shade for a child, he thought. Not even a child's foot, or hand. A hot blue dome of cloudless sky held them under its lens. The boy did not belong here.

That tree is where he was found.

The boy had walked past the branching cholla that crowd the trail, reaching for hikers' skin with their barbed spines. He had passed a towering saguaro cactus, icon of the Sonoran Desert, whose silhouette—the largest shade on the hillside—was cast across his path like a gate, warning him not to go. He was a Boy Scout; he had good manners. He did what adults asked.

"This is our natural disaster," his father would say later, to reporters gathered in the parking lot, glancing up from their microphones to the dry peak behind him. "The heat, here in Arizona, is our natural disaster."

The paramedic started an IV in the boy's arm and began pushing cooled fluid through it. His partners stripped the boy's clothes in a flurry, pulled bags of ice from their packs and stuffed them into his armpits and groin and neck. The men looked around every few seconds, searching for shade they knew wasn't there. One of them was at the boy's head, intubating him, so they could keep him breathing with an inflatable bag. A drop of sweat fell from the man's face, onto the boy's. That stays with him; he doesn't know why.

He kept talking to the unconscious boy, telling him help was coming. A thumping dot in the sky was getting closer: the helicopter that would carry him to Phoenix Children's Hospital. The men raised the boy on the gurney and surrounded him for the trip downhill, jogging alongside, trying to hold everything in place. A bend brought the parking lot into view again, small and distant below. That was where the helicopter would land; that was the spot they had to reach.

He saw a car driving south on Dove Valley Road, a dark thin line through the bleached expanse of desert. The car's glare moved across the flat like a targeting beacon. It slowed at the parking lot entrance.

Two hours earlier, on the afternoon of July 22, 2016, Cody Flom had come down that lonely road with his mother's boyfriend, the guy who pointed them up this hill. They had turned into the Apache Wash trailhead.

It was 112 degrees.

Extremes

I flew to Phoenix the following spring to meet with Cody's parents and others involved in his case. My plane crossed the northern part of the city, not far from the mountain where he was found. As we descended I saw our shadow pass over the dry ground, pointing like a cursor to its past: to centuries of unbroken sun; to water's terrible absence, and its fleeting and forceful hand. Rivers of dust flowed across the plain, emptying into arid catchments; thirsty scrub huddled on their edge. Heat is invisible, except it's not: you can see what it has done.

It was heat I had come to study, though not its effect on terrain. The fundamental problem of global warming is also a profound threat to children's health. Everywhere is getting hotter; heat waves sweep the

Earth with increasing, alarming intensity. Preparing for them is a looming public health challenge for cities all over the planet. But the already-hot regions of the world, including the American Southwest, are feeling the impact first, and most.

I had flown from Reno, the fastest-warming city in the country. But northwestern Nevada had not started out as a very hot place, and most of the parents in my pediatric clinic—even most of the other physicians I knew—weren't aware of our increasing risk. In their own childhoods, whole summers had passed without a 100-degree day; now half of July could cross that threshold. I had started warning parents that children needed to be watched closely in the heat, and given frequent shade breaks with plenty of water. Summers were changing.

Doctors in Phoenix had been dealing with heat far longer. No American city has more "extreme heat" days, above 99 degrees; the average summer high is 105, and several weeks a year are over 110. Phoenix *knows* heat—coping with it is a big part of the city's culture and habits, and a point of local pride.

Its residents are facing something different now.

Imagine a bell-shaped curve, showing the number of days at various temperatures in any one place. At the extreme right are very hot days; at the extreme left, very cold. The vast majority of days are in the middle, moderate and tolerable. This is the world we have known. Now imagine that bell curve being pushed significantly to the right. Many more hot days, many fewer cold ones, and average days that are significantly warmer. That is the world we are creating.

Phoenix's bell curve was already to the right of most cities; as it gets pushed further rightward, the number of hot days, and the hottest temperature possible, rises. The five-year period leading up to Cody's hike was the hottest Phoenix had yet seen. It was capped, three days before he headed up the mountain, by startling news: the already-blazing city was the second-fastest-warming in the country, right behind Reno.

If I wanted to understand heat's growing risk to the children I care for, Phoenix was the place to go.

Officially, about 67,500 people seek medical care for heat exhaustion and heatstroke in the US annually, and roughly 700 die—more than are killed, in most years, by flooding, hurricanes, and tornadoes. But tracking heat illness is challenging, and those tallies almost certainly understate its actual impact. Babies and small children, the elderly, those who work or recreate outdoors, and low-income people are especially at risk.

Arizonans die from heat at up to 7 times the rate of the rest of the country. The summer of 2016 was particularly brutal. In a single weekend in June—just a month before Cody would ride down Dove Valley Road—six people, all using the region's popular trails, were killed by some of the highest temperatures ever seen in Tucson, Phoenix, and Yuma. In Maricopa County alone, 154 people died that summer from heat-related illness—then the largest toll in its history.

Perhaps not coincidentally, it was, at the time, the hottest year ever recorded on Earth.

In a place where summers can already reach 122 degrees, rapid warming is pushing Phoenix toward the limits of survival. The city "could regularly hit the 130s . . . by the second half of this century," according to University of Arizona climate scientist Jonathan Overpeck. In temperatures that high, just going outside will be a risk to life.

Children won't have to live in Phoenix to be affected, as events since Cody's hike have shown. In June 2021, the world was stunned when an unprecedented heat wave of Arizona-like temperatures struck normally cool Oregon, Washington, and British Columbia, killing hundreds of people in cities unprepared for such extremes. The small town of Lytton, BC, hit 121 degrees, and the next day burned to the ground in a wildfire.

Though it may seem like Earth's future warming is inevitable, a swift transition to clean energy would quickly stop the thermometer's upward climb. Yet without that rapid and widespread effort to alter our path, a child born today will see Earth's temperature rise significantly above what we are experiencing now. By 2100, when that child is elderly, tens of thousands of Americans could be killed every year by heat. The world our children and grandchildren inherit may be so dangerously hot, one researcher said, that "we will become prisoners in our own homes."

Of course, he really meant *they* will be. We'll be gone, having left them a future we saw coming but chose not to prevent.

I read temperature projections in my hotel room the first night I was in Phoenix. Shaken, I turned off my computer and headed out for a walk. Even though it was after dark, and even though it was only May, heat radiated from every surface downtown, and took my breath away when I stepped outside.

Over the street ahead of me, a billboard claimed that only one soft drink could cool Arizona. A mother, pushing her baby in a stroller, hurried beneath it to a bus stop; a little girl in shorts ran behind her, trying to keep up. I watched the young woman reach the waiting bus, unstrap the baby, and lift him onto her hip, then use her free hand to fold the stroller and search her purse. The little girl, breathing hard, looked up at the driver through the open doors of the bus. Her brother's bare legs straddled their mother's waist; as he fussed, his mother turned and comforted him. *No pasa nada, mi amor, no llores.* It's nothing, my love, don't cry.

I jogged to lend a hand, but before I could get to her she had found the pass and they were climbing the steps. The bus pulled away, leaving behind a cloud of exhaust that dissolved into the hot air. I stood on the empty sidewalk and gazed at a slice of nearly starless sky, framed by the dark office towers on either side of the street. Though their workers had gone home, the buildings' signs stayed lit and their air conditioners hummed. The city was untroubled. Yet all around me its exhaled fumes were invisible and rising.

Back in my hotel, I reread the news release from the prior summer about Phoenix's rate of warming. It had not mentioned those with the most to lose. Media coverage of the story had been scarce; what it meant for the youngest, nonexistent. In neighborhoods throughout the city that week, children had retreated into their homes as temperatures climbed.

One of those children was Cody Flom. On July 19, the day after climate scientists pointed to Phoenix and the other fastest-warming cities on their list, Cody had returned from a California beach vacation with his father, stepmother, and one of his friends. He had another month of summer break before starting seventh grade. Cody liked Minecraft, and swimming; he was sweet to his three-year-old niece, and especially close to his grandmother. His favorite color was hot pink. "I think he liked the attention it brought," his father told me when we met. On cooler nights, he'd go to the park and run passing routes while his dad tossed the football.

On July 20, he went to his mother's house. Two days later, the National Weather Service issued an Excessive Heat Warning for southern Arizona, indicating "extremely dangerous heat conditions" that risk serious illness or death. That afternoon, in the middle of a three-day heat wave, Cody was told to get out of the house and go to the car: they were going hiking. Just a mile out and a mile back, his mother's boyfriend said, and threw two bottles of water in the back seat.

"He was a good kid," his father said, staring at the table between us. "He'd have trusted an adult."

The day after my arrival, I sat in a Phoenix neighborhood of stuccoed houses and succulent yards where the signs of children were everywhere: playgrounds, birthday balloons tied to a mailbox, a minivan with a stick-figure family on the rear window. At 100 degrees, the street was as silent as a Martian outpost. Swings hung motionless on their chains. When a rare breeze gusted, I was grateful for its cooling but also its sound; for the relief from stillness. After about an hour a school-age girl, dressed in cutoffs and a T-shirt, left her front door and walked hurriedly to the next house, where she was quickly let inside.

"It was a spring without voices," Rachel Carson wrote in 1962, imagining a future where environmental calamity had wiped out the routine music of nature and childhood. In peak heat, Phoenix's tidy grids of residential streets, sprawling outward from downtown, are eerily empty. I saw in that neighborhood a window into my patients' futures, when extreme hot temperatures have become the norm, when children will not be able to play safely outside for any length of time. Instead of Carson's *Silent Spring*, many cities of the world may soon have silent summers.

"A grim specter has crept upon us almost unnoticed," warned Carson, "and this imagined tragedy could easily become a stark reality we all shall know."

The Phoenix area is home to 4.8 million people, and still growing. Many in the region's universities, government, and medical community are working on ways to adapt to rising temperatures, reduce urban heat, and prevent illness. But there is only so much the city can do if the broader world does not quickly reduce the greenhouse gases causing Earth's temperatures to rise. Without fast action, America's fifth-largest city may be virtually uninhabitable within the lives of its youngest residents.

It's an ironic future for a town whose growth was once fueled by people seeking a healthier environment—and whose main attraction, in the past, was its weather.

"Nature's Sanatorium"

In 1919, a young California couple packed up their two children and all their belongings and set out in a touring car for the desert. Marguerite and Albert Colley were not wealthy: he was a farmer who had done side work in the gold fields; she was a practical nurse and social worker. But they were desperate to help their nine-year-old son, Robert, who struggled with recurrent bouts of cough, wheezing, and shortness of breath—and Arizona, everyone knew, was the place to go.

The Colleys were "healthseekers," or "lungers," people who moved to Arizona because they believed its warm, dry climate could relieve, or even cure, lung disease—in this case, their son's. Tens of thousands of healthseekers went to the Southwest between 1880 and 1930, most suffering from "consumption," a catch-all term for a variety of respiratory illnesses, but usually referring to the most dreaded: tuberculosis.

Tuberculosis was one of the era's most pernicious diagnoses, a shroud darkening every aspect of human experience. In the nineteenth century the disease killed more people in America and Europe, by far, than any other cause—a wave of death dubbed the White Plague for the pallor of its victims. But another respiratory infection may have triggered the Colleys' move. In 1918–1919, the Spanish influenza pandemic killed 675,000 Americans, nearly 3 times the per capita death toll of COVID-19. The Colleys had multiple reasons to worry about Robert's already-struggling lungs, and to seek the touted protections of Arizona's climate.

I came across the Colleys' story in a small historical society in the north Phoenix neighborhood of Sunnyslope, where many of the first healthseekers had settled. "Most of us can't fathom it," one of the elderly volunteers said, as she brought me a thick file of century-old letters and photographs. "All those thousands of people moving cross-country for their health." I thanked her and didn't mention that, given predictions about Phoenix's temperatures, I could fathom such migration a little too well.

The idea that drove the Colleys was not new. Hippocrates, 2,500 years before, had argued that illness or wellness arose from "air, water, and place"—the climate in which a person lives. His theories resurfaced in the eighteenth century when a group of London physicians, captivated by early thermometers, barometers, and other weather gauges, began recording weather data in their notes about patients and epidemics. They believed that many of the maladies they saw in their clinics were caused by variations in climate, and that this could be proved through meticulous observation. Over decades, through their analyses of wind, rain, temperature, humidity, and atmospheric pressure, the science of meteorology was born. But it arose from the practice of medicine, and many of the first meteorologists were doctors.

During the nineteenth century, as the White Plague raged, "medical meteorology" continued to shape doctors' thinking. In 1885 it led to the establishment of the first American "sanatorium" in upstate New York. In this spa-like facility, TB patients lived in tents to maximize their exposure to fresh air—the key, doctors thought, to clearing phlegm-filled lungs. Sanatoriums spread rapidly, especially in warmer regions where the sick could be outside year-round, and by the 1890s, "climatology"—the idea that certain levels of humidity, altitude, and temperature foster good health—was widely accepted. (Of course, the word has a very different meaning today.) Many physicians ascribed to it, routinely advising cross-country moves to their consumptive patients.

By the time the Colleys reached Phoenix a quarter century later, Arizona was the top destination for tuberculosis patients seeking care at sanatoriums. "Nature creates and maintains; she must therefore be able to cure and rebuild," explained a 1910 brochure for one Phoenix sanatorium, summarizing the movement's philosophy. The desert was touted as "most curative for those afflictions which depend for relief largely, if not entirely, upon climatic conditions." As testimonials spread, Arizona became known as "nature's sanatorium," an idea the state's promoters did nothing to discourage.

And people did sometimes get better—though for most, improvement came simply from escaping the overcrowded, polluted conditions of nineteenth-century cities. As I sat in the historical society, studying the photos of desperate migrants streaming into the desert wilderness, it struck me that their doctors were wrong in the specifics, but right about a bigger point—one that physicians largely forgot in the century that followed. We don't look to the sky for answers now. Yet our health is still inseparable from the air, water, and place that surround us.

Marguerite Colley and her family settled near North Mountain, not far from the small building where I would later review the records of her life. Around her, thousands of healthseekers had flooded a new state that had little ability to care for them. Most were poor, sick, and alone, having left friends and family far behind. Unable to afford the high-end sanatoriums, desperate consumptives struggled to get by, dotting the landscape with their tents and shacks.

Marguerite began caring for their needs. By the 1920s she was running a small medical clinic, getting advice from distant doctors over the just-laid phone lines. In one photograph, she looks serious in a white nurse's uniform, standing next to the clinic's new ambulance. A medical bag sits at her feet; a young girl, strapped to the ambulance's gurney, looks grimly into the camera from the past.

There is another photograph of Marguerite. She is with her children at a local grocer's. Robert Colley stands next to his mother. His health improved in the desert; in her preserved writings, Marguerite never mentioned it again. Robert grew up, worked in real estate, had a family, and died at the age of ninety-five. Apart from its length, an unremarkable life.

But Robert Colley's childhood asthma had a "butterfly effect"—it was a small detail that left a permanent mark on the city of Phoenix, and indirectly saved the life of a boy he never met.

In the 1930s, a wealthy migrant to Sunnyslope, grateful for his wife's recovery from consumption, donated to Marguerite's clinic. His funds enabled her to expand the clinic into a small hospital—eventually the John C. Lincoln Medical Center of north Phoenix, named for its original benefactor.

Almost a century after Marguerite brought her son to the desert for his health, the hospital she founded would save another woman's son from that same desert. Joey Azuela was nearly killed by heat—by the climate, once sought after—in the spot where the TB sufferers' tents had stood.

Joey

When Alicia Andazola heard about Cody Flom, her heart sank. "That was almost us," she tells me, her voice shaking. "It brought me to tears." She had called the local news. Maybe some good can come of what we went through, she thought; maybe we can warn other parents.

On August 1, 2015, her son Joey had, like Cody, gone hiking on a hot day. He had also suffered severe consequences; in fact, Joey's case was almost identical to Cody's. But their stories differed in one critical factor: speed.

The day of his hike, Joey was excited about the weeks ahead. He had just registered for his freshman year at high school and was already practicing with the football team. Trying to get in shape, the fourteen-year-old had also been training with his father. That afternoon they headed to North Mountain Park, a set of straight-up, straight-down trails in the Sunnyslope neighborhood of Phoenix.

It was 103 degrees—not bad, by local standards—and the pair had hiked that day's route before. Neither of them was alarmed when, shortly after starting out, Joey complained of not feeling well and vomited. He rested for a minute, then continued to climb. About halfway up the mountain, he and his dad ran out of water. It wasn't going to be a long hike, they reasoned, only an hour; they'd drink plenty when they got back to the car.

Joey was tired but pushed himself to the top. Like most desert trails in the area, this one offers almost no shade. When they reached the peak he complained of feeling dizzy, but the hike was meant to be a workout, so they were quickly on their way again.

"My dad thought because he was fine, I was fine," Joey tells me, "because we had the same amount of water." The teenager remembers his legs and body felt hot. "On the way down, I felt weird, like I was floating almost," he said. The rest of the return trip is blank, except for the end: he saw his dad's car in the parking lot, and thought, for a split second, he had made it.

Joey collapsed onto the hot asphalt. He was wearing shorts and a T-shirt; when his skin hit the pavement it was like landing on a stove. His arms and legs were severely burned. His father tried to drag him to shade, tearing the burns open and embedding them with dirt and pieces of asphalt. One arm was so badly bruised from his fall the ER doctors would think, mistakenly, that it was broken.

As Alicia rushed down the hallway of John C. Lincoln hospital, she saw Joey's father sitting on the emergency room floor, crying. When she'd first gotten his call, she was confused. Why did Joey need an ambulance, when he had just passed out? Now she was beginning to understand: this was something much more serious.

She stepped into the room. Her son lay on the bed, stripped of clothes, unconscious, and covered in ice. The small space was filled by chaos. "It was 'all hands on deck' . . . he was just surrounded by people," she recalls, each of them in motion, frantically trying to save his life. A nurse was scooping more ice onto him; as Alicia stepped toward his bed, she slipped on a piece that had tumbled to the floor. At some point someone, a doctor maybe, explained that he was in a coma, that a ventilator was breathing for him, and that the other machine he was hooked to was extracting his blood through a large tube, cooling it, and returning it to his body.

On arrival in the ER, his core temperature rose to nearly 107 degrees. Joey had suffered heatstroke.

"It gives me goose bumps thinking about it," she says, recounting all the circumstances that saved his life. We are sitting at her kitchen table, almost two years later, several inches of medical records stacked neatly in front of us. "He was already at the bottom of the mountain when he collapsed, so they didn't have to hike up to get him, and that level-one trauma center was so close, he was able to get treatment right away."

John C. Lincoln Medical Center—the hospital that grew from Marguerite Colley's clinic—is just over a mile from where Joey fell. Its Level 1 designation means it can provide the highest level of trauma care available. Joey's good luck saved him from the bad; had he and his father chosen any other park, he likely would not have survived.

After a few hours, with his temperature down to 98–99 degrees, he was transferred by helicopter to Phoenix Children's Hospital—the same hospital Cody would be brought to the following summer.

Even though Joey's temperature had returned to normal, major damage had been done. When he arrived in the pediatric intensive care unit, he had a seizure. The records from Phoenix Children's tell a grim story: not just his seizing, but every organ in his body was injured. Heart and liver enzymes were elevated, indicating that those organs' cells had been damaged. He was on a ventilator due to respiratory failure and unconsciousness. The lining of his intestines died and sloughed off. He experienced massive muscle breakdown, called *rhabdomyolysis*; the toxins released by his dying muscles caused kidney failure, which took months to resolve. Doctors warned his family that he might never be the same. He had to learn to write, talk, and walk again, and missed his entire first semester of high school.

The summer of 2016 was not easy for Alicia.

"I mean, I counted . . . I counted and I kept track," she tells me, referencing Arizona's heat-related illnesses, rescues, and deaths that summer. But Cody's case hit especially hard. "That one really tore me up."

Shortly after Cody's story was in the news, camera crews came to Alicia and Joey's home. Joey told them what he could remember and bared his scarred arms and legs, which had taken nearly ten months, and many painful dressing changes, to heal. Alicia shared photos of his worst days, tethered to life by multiple machines. *Teen spent year recovering after near-death hike*, one caption read. *Brush with death*, said another. Alicia tried to warn others, especially parents. "The summer in Arizona, the heat, it kills," she said.

I look at Joey, sitting on the couch, now a tall, muscular adolescent, nearly full-grown. Apart from the scars, he is intact, a miracle to his mother—who needs no reminding that, despite everything, they were lucky.

The Body's Furnace

Later, I would remember Alicia's expression as she looked through Joey's medical records: like she couldn't believe what had happened to him, even as she described it in detail. Though I didn't say anything at the time, I knew there was another layer to his story, hidden in those records. His temperature had reached 107 degrees on a day that was only 103. His body had gotten hotter than the desert. How was that possible? And what did it tell us about the dangers of an ever-warmer world?

The answer lay in Joey's own *metabolism*: the chemical reactions that were happening in every cell of his body. These reactions release heat; they are like a furnace that's always on, and the reason a living body is warm. To keep him cool, his brain and nerves, heart and blood vessels, skin and sweat glands must work together to shift this heat to his surroundings. But when the outside world is too hot, all these cooling tricks eventually fail, trapping metabolic heat and driving body temperature upward—as happened on the day of his hike.

When that afternoon began, his metabolism was performing its usual work. Before taking a single step, while sitting in his father's air-conditioned car and lacing up his boots, the meal he had eaten before the hike was being digested: broken down into molecules, like glucose and protein, his body could use. As the food's chemical bonds were broken, energy was released, mostly as heat. But some of that energy was used to form new bonds in a chemical called *adenosine triphosphate*, or ATP.

An ATP molecule is like a rail car carrying energy to anywhere in the body, for anything it needs. As Joey began to climb uphill, the energy he needed to contract his heart and leg muscles came from breaking apart ATP, freeing the energy stored in its chemical bonds. Once again, most of this unleashed energy became heat; in fact, only about a quarter of it was used to propel him up the mountain. (This comes in handy at the opposite end of the thermometer; when we shiver in the cold, we're warming ourselves by breaking apart ATP.)

If he didn't have a way to dissipate all this heat, Joey's internal organs would rapidly become too warm. But that doesn't happen because his blood, as it circulates, continuously transfers heat from the core of his body to his much cooler, 91-degree skin, where it radiates to the air. The cooled-off blood then returns to his core, keeping his organs close to 98.6. On a moderate-temperature day, about 5 to 10 percent of the blood pumped by our hearts goes to our skin.

When Joey stepped from the car into a hot and sunny August day, though, his skin temperature rose, decreasing his ability to cool his core, which then began warming from the work of his muscles. Special nerves in Joey's body sensed his rising temperature and sent a message to a part of his brain called the *hypothalamus*—specifically, the *preoptic area of the anterior hypothalamus*, or POAH, "the body's thermostat." The POAH responded by sending signals outward to his skin, muscles, and other organs.

These outgoing nerve signals triggered dilation of blood vessels in Joey's skin, raising its temperature and releasing more heat. Blood flow to his skin had now jumped to about 33 percent of cardiac output.

But on that very hot day, increased blood flow couldn't overcome the fact that the environment was a higher temperature than his skin. He continued to warm. His body's only option at that point was to cool through evaporation.

So the POAH then told millions of glands in his skin to produce sweat. Sweating increased as his temperature rose, both through *apocrine glands*, which produce odorous sweat mostly in the armpits, and *eccrine glands*, which emit odorless sweat over most of the body. The eccrine glands are more important for heat regulation because their sweat has less salt and evaporates more easily.

To continue sweating, Joey needed fluid. But here he ran into a basic problem of math and biology. Exercising on a hot day, he was losing about 2 liters per hour through evaporation—not just from sweat, but also from the mucous membranes of his mouth, nose, and lungs as he breathed. Though he'd been drinking since hitting the trail, he was only able to absorb about 0.5 liter per hour from his digestive tract—a quarter of what he was losing. At this rate, he was dehydrated within an hour after leaving the car, and the situation got worse the longer he hiked.

Joey and his dad had not taken enough water; symptoms of *heat exhaustion* began early in their climb. He felt weak and tired, his head hurt, he became nauseated and vomited. He was sweating profusely, and his muscles were starting to cramp. By the time he left the peak, his temperature was probably already above 102 degrees.

The fluid needed to make sweat came from his blood: from capillaries in the skin that feed the sweat glands. But as he continued sweating, and didn't drink, his blood volume and blood pressure dropped. In response, to keep his organs supplied with oxygen, his heart beat faster. Receptors in Joey's brain signaled the sensation of thirst, and hormones told his kidneys to stop making urine. His blood became more concentrated, thus higher in salt, increasing his risk of seizure. He became fuzzy-headed as blood flow to the brain dropped; he remembers little about his descent.

Had he been able to rest in cool shade and drink, these problems might have been resolved. But he was trying to get a good workout, water and shade were nonexistent, and he wasn't reasoning well. He kept going. Heat exhaustion was progressing to *heatstroke*, a life-threatening emergency.

Blood started to pool in his dilated skin veins, forcing Joey's body to make hard choices. It routed less blood to internal organs and skin, more to muscles and brain—the most important organs for surviving this situation. His heart rhythm became irregular because of low blood flow, heat damage to the heart muscle itself, and high levels of the blood salts that regulate heartbeat. Blood flow to the intestines dropped—the reason their lining died and later sloughed off—meaning that even if water had been available, absorbing it would have been difficult. His too-hot muscles were breaking down, releasing the toxic proteins that damaged his kidneys. At this point his core temperature likely rose to 104 degrees due to the total loss of convection from his skin, which had stopped sweating and had become hot, dry, and red.

If Joey had not been rescued and cooled quickly, his brain and other organs would have suffered irreversible damage and he would not have survived. As it was, his temperature reached 107 in the hospital; any prolonged period over 105 is fatal. In heatstroke treatment, the speed of cooling is everything—as the paramedics who tried to save Cody Flom knew, too well.

The Fiercely Protected Range

"Humans live their entire lives," physiologist W. Larry Kenney has noted, "within a very small, fiercely protected range of internal body temperatures." Hot blood and cold hearts are great metaphors, but in truth, when we are healthy, our body's core rarely strays more than a degree from 98.6. We stay in this range through a combination of conscious decisions, such as putting on a sweater or taking it off, and involuntary acts like sweating.

It is eye-opening to observe, for a single day, how much we do to stay within our comfort zone. From our clothing decision in the morning, to how high or low we set the thermostat, whether we drink cold soda or hot chocolate, stay inside or go out, sit in sun or shade, go to the pool or sit by a fire, open a window or lower a blind—we are constantly reacting to the temperature of our world.

And our body is adjusting all the time, without any direction from us. When we sweat, shiver, or feel the sensations of hot and cold and thirst, we don't ponder all the amazing things that just happened—the genetic code passed down over several million years of adapting to the weather on this planet. Or how these reactions depend on the remarkable properties of our own skin, where our body meets the surrounding air. It is at this boundary that messages exchanged with the brain allow us to assess and cope with the temperature fluxes of our world—again linking the atmosphere's health with our own.

But all this begs the question: *Why* do we need to stay within a certain temperature range? Why does it matter that our internal environment is thermally stable, or that Joey's body temperature climbed so far above normal?

The answer lies in the chemical reactions at the heart of life itself. These reactions—such as the building and breaking apart of ATP, which we do millions of times a day—are affected by temperature. Our metabolism functions best in its normal temperature range; if it gets far outside that range, our bodies stop functioning at the most basic level. If temperatures are extremely high, the chemical bonds that form our organs and other body structures come apart: proteins unravel, cells rupture, and tissue starts dying. This is what happened to Joey's heart, muscles, intestines, liver, kidneys, skin, and brain.

As I studied Joey's story, though, a mystery began to nag at me. He should have been acclimated to heat. When people are exposed, gradually but steadily, to increasingly high temperatures, their responses start improving. They sweat sooner, and more; their hearts get stronger, better at handling the demands of cooling the body. Athletic coaches in hot climates and military commanders in the Middle East are very familiar with this process of *acclimation*. It is the reason that tourists in Arizona face higher risk of heat illness than locals. We don't respond well to big temperature variations from normal. But Joey lives in Phoenix and had been training for football.

His case illustrates that acclimation improves our responses but cannot change the basic chemistry of life. There are hard physical limits to what we can survive. Joey ultimately reached those limits; exercising on a hot day, especially in the absence of water to replenish his sweating, overwhelmed his ability to compensate—a situation known as *exertional heatstroke*.

Many public health researchers, looking to the future, pin some of their hope on acclimation. It will help, and we have to take advantage of every tool we have. But at some point, between the atmosphere of Earth we have known and the atmosphere of, say, Venus—where carbon dioxide is so high that surface temperatures can melt lead—lies a threshold of widespread heat illness, and worse.

For children, that threshold is lower than for their parents.

Children Are Different (Again)

"Kids are vulnerable," she says. "More vulnerable than you or me. And they don't have the experience or knowledge to know when to stop."

Jennifer Vanos leans forward, clasping her hands together on the table. I had just told her about Cody and Joey. "A lot of the work I'm doing is trying to understand exactly what's happening to a child in conditions like that," she says.

Dr. Vanos is a scientist who was then at the University of California, San Diego. We were sitting in her new, spare office, in a modern building on a cliff above the beach. Behind her, through the window, was the Pacific Ocean; a squadron of pelicans flew by in the distance. These were not the typical drab digs of most public health research. But Vanos's work was something new: she had a joint appointment at the Scripps Institution of Oceanography and the UC San Diego School of Medicine.

Like a growing number of medical schools, UCSD had recognized that the climate crisis is not strictly an environmental problem. Young doctors must be prepared for illnesses that are increasing as the world warms. The school had assembled a multidisciplinary team of researchers to study these growing health risks and, hopefully, to develop adaptations that can protect us. Vanos had just joined the faculty.

Her interest is *heat physiology*—how and why the body warms and cools. Much of our understanding of how heat affects us comes from experiments on adults: military men, outdoor workers, athletes, and the elderly, who have the highest rates of heat illness and death. Only a small percentage of heat data comes from children. Vanos aims to correct that. She is a *biometeorologist*, a scientist who measures the impacts of weather and climate on living things—in her case, children. She had done research in Phoenix.

Children's bodies are different, she explained, in ways that make them significantly more susceptible to heat. First, they have more surface area, relative to their weight, than adults. Pound for pound, a newborn has 3 times the surface area of his parent; a school-age child about 1.5 times more. That's why babies and young children gain or lose heat more quickly when placed in hot or cold environments.

Children also produce more metabolic heat per unit of weight, even at rest. Parents in my clinic often joke about wanting to bottle their kids' energy; that energy comes from a faster metabolism. It's why their movements, breathing, and pulse are quicker than ours. But the most important reason they generate more heat, relative to their size, is that they are constantly growing, an intense metabolic process unique to the young.

Like all of us, they need to move this metabolic heat to the skin to radiate to the outside world. But compared to adults, babies and small children have less blood with which to dissipate it. And their heart muscle is not as strong: they can't increase the *force* of the heartbeat as much as the *rate*. A faster rate gives the heart less time to fill between strokes. So as the body heats it may not get enough oxygenated blood, causing tissue damage. This is especially a problem for newborns, whose resting heart rate is already 140 to 160 beats per minute.

In heat or with exercise, adults sweat sooner, and more. Toddlers' sweat glands are smaller and produce less than half the sweat of adults, and their bodies acclimate to hot weather more slowly. This is even more true for girls: males sweat more than females throughout life.

All these differences mean that babies and young children heat faster than their mothers and fathers. The gap narrows as they grow and age; by fourth to sixth grade, most kids have roughly the same ability to exercise in heat as adults. But they continue to face more risk for developmental and behavioral reasons. Children and even teenagers rely on their parents to notice if the weather is too hot, offer water, or move to shade. If the adults themselves don't have symptoms, they may not realize kids are getting into trouble. Both Joey and Cody were with men who, in the exact same circumstances, did not get ill, and missed that the boys' lives were in danger.

Jenni Vanos is trying to quantify all these risks—and protect kids' ability to venture outdoors.

Protecting Children—and Childhood

Vanos took me to her laboratory to show me the science-fiction gadgets she has children wear as they play. "We're trying to get data at the scale at which kids are experiencing it," she says, handing me a small stainless-steel disc that snaps onto a child's belt loop. "This one measures temperature and humidity." Putting monitors on kids enables her to record the "microclimates" they move through, such as a shady wooded area or a hot basketball court, and see how their bodies respond.

Her real-life laboratory: the city playground, where Vanos has shown that small details make a huge difference in kids' safety. She found that Phoenix playgrounds often sit idle in summer for good reason: the surfaces of their slides, swings, and Astroturf can reach 180 degrees. That's enough to melt kids' shoes—and hotter than the asphalt that burned Joey Azuela.

With an eye to what's coming, Vanos lobbies planners around the country to include shade sails and trees in their playground designs, and to use more wood in place of metal, plastic, and concrete. What if playgrounds were not hot, glaring, and uncomfortable, she asks, but a place of cool refuge, where kids could see birds and play with bugs and leaves?

In some ways, Vanos's work aims to protect the very *idea* of childhood in a hotter world. Studies have shown that the more time children spend in nature, the better their mental and physical health. The more they become stuck in front of screens because the outside world is too hot, the further they will travel down a path to obesity, diabetes, heart disease, and depression.

Her mission has a dual purpose. Hot playground equipment also adds to the urban heat island effect. Cities can be 15 to 20 degrees warmer than surrounding areas because of heat radiating from buildings and roads—the oven-like quality I had noticed in downtown Phoenix. Low-income neighborhoods are significantly hotter because they typically have the fewest trees and green spaces. Climate change is increasing this urban heat risk to children; "natural playgrounds" could help reduce it. When it comes to protecting kids from future heat, Vanos is advocating for a simple step that would improve every community in America.

Heat's Hidden Toll

The doctors who treated Joey and Cody had no trouble naming what made the boys ill. Heatstroke was obvious from their body temperature and the circumstances. But in many other cases, heat's role is hidden.

Imagine, for example, that the little girl I saw running to the bus stop on my first night in Phoenix had complained of headache and vomited after she got home. Those are common symptoms that I see often in my clinic, with a long list of possible causes. If heat were the culprit in her case, how would her mother know?

And what if their air conditioner wasn't working, and her bedroom stayed hot all night? Cooling off in the evening, it turns out, is very important to our health; it reduces the ill effects of hot days. Unfortunately, increasing humidity and urbanization are driving up nighttime "daily minimum" temperatures even faster than "daily maximum" temperatures in many places, including Phoenix. (In fact, the city experienced its highest-ever overnight low temperature, 97 degrees, in 2023.) What if by the next morning the little girl was weak, feverish, and still vomiting? If her mother took her to the pediatrician at the local health clinic, would the doctor recognize that she was suffering from heat exhaustion, not food poisoning or any of a dozen other possible diagnoses?

There's no easy test that can tell him. He would likely end up prescribing fluids and antinausea medication to treat her symptoms, and when entering a medical diagnosis in the clinic computer, choose *Vomiting, unspecified* without naming a cause. This is acceptable practice, and she would almost certainly improve. But if he doesn't select *Heat exhaustion* as her diagnosis—or if her mother never takes her to the doctor at all—her case will not appear in Maricopa County's annual tally of heat-related illness.

Variations of this scene play out thousands of times across the country every year. Heat's effects are often stealthy and nonspecific, and tabulating them is notoriously difficult. For this reason, public health officials assume that their counts of heat illness and death significantly understate reality. This may be especially true for infants and young children, whose greater susceptibility to heat isn't on every doctor's radar. From 2006 to 2010, physicians ascribed fewer than 300 pediatric deaths in the US to heat; less than 3 percent of recognized heat-caused hospitalizations are for children.

So how can we know the number of people actually getting sick or dying from hot weather? By comparing historical rates of illness and death for a given date to those occurring during a heat wave. "Excess" deaths or illnesses, above the norm, are attributed to high temperatures. Using these methods, we know that a three-day heat wave can increase all-ages, all-causes mortality by 19 percent; the higher the temperature or the longer it lasts, the worse the toll. And we've found that heat may kill more than 5,600 Americans every year—8 times the official CDC tally.

Several large analyses of this type have shown that babies and toddlers have much higher heat mortality than official statistics would suggest—in fact, the highest rate after the elderly. The younger the child, the greater heat's threat: infants under one year of age, particularly newborns, are most at risk. Among all age groups, they are expected to suffer the greatest increase in heat mortality as the world warms, especially in areas with low rates of air-conditioning.

Using these methods, abnormally hot days have also been linked to a wide range of health problems in both children and adults. If the little girl at the bus stop had a grandfather with congestive heart failure, for example, his condition could worsen because of the demands heat puts on the heart. Her baby brother would be more in danger of SIDS.

If her mother were pregnant, she would face higher risk of heat illness and life-threatening pregnancy complications. She would also be more likely to lose the pregnancy or deliver prematurely, with all the challenges that entails for her baby. If the heat wave occurred during the first trimester, her infant would have greater odds of birth defects, including *spina bifida*, a condition in which the spine doesn't finish forming. And exposure to heat in the second and third trimesters can lead to low birth weight, which has major, lifelong health consequences. Low-birth-weight babies grow into adults with higher rates of cardiovascular and renal disease and diabetes. One apparent reason is that prenatal stress not only impairs fetal growth but also alters the way certain genes are expressed, making these conditions more likely. It's another powerful example of how an environmental insult to a young body can impact the rest of that person's life.

As we saw in the last chapter, air pollution can trigger many of these same complications. Heat waves worsen air quality because they increase ozone levels and often prompt wildfires. The heat risks to pregnant women and their babies are thus compounded by the dangers of air pollution, especially in low-income areas that are hotter and have more traffic.

In addition to vomiting and headache, the little girl would face a range of threats, depending on her overall health. Children with asthma sometimes cough and wheeze more during and right after heat waves, because of the bad air. Diabetic children have higher blood sugars; epileptics, more seizures. Kidney injuries increase because of dehydration, especially in obese children or those with disorders that impair sweating, like cystic fibrosis. For all these reasons, pediatric emergency room visits often jump in very hot weather—though again, heat is rarely blamed.

But some ER doctors have definitely noticed one tragic problem that spikes when the thermometer climbs.

Heat and Human Behavior

One summer night toward the end of my residency in Sacramento, I was working an overnight shift in the pediatric ER when a two-year-old was brought in by ambulance. He and his mother had been attacked by a neighbor who'd become enraged when the air-conditioning in their apartment building failed and the toddler, unable to sleep in the heat, kept crying. Some of the paramedics and police who had been at the scene lingered in the ER hallway for word of the baby's fate; I saw a giant of a man in fire department suspenders, sobbing. When the nurses couldn't get an IV started, I held the boy's tiny calf in my left hand and with my right pushed an intraosseous line into his tibia, to give him fluids through his bone marrow. I am accustomed to small bodies but the size of his, relative to the violence committed against it, is an image I've never been able to erase.

Not until several years later, when working on this book, did I search temperature records and discover that the man's attack had occurred during a burst of abnormally hot days: a heat wave. It gave me a partial explanation for the inexplicable, but no comfort. A growing field of research shows that heat doesn't just affect our bodies, but our minds—and with each uptick in the thermometer, more women and children are mistreated.

Intuitively, we all understand why. When our bodies get hot, we get angry more easily and become less rational; our threshold for violence drops. Experiments done by psychologists in the 1970s showed that a person experiencing an annoying event—such as, perhaps, a baby crying next door—was more likely to become aggressive if temperatures were higher.

Researchers have now identified multiple ways this plays out in everyday life. In hot weather, drivers are more likely to honk at one another. Baseball pitchers hit batters more often. Violent crime rates go up.

For children, the most serious effects are in the home. Adult depression and other mental health problems are exacerbated by heat waves, decreasing parents' ability to care for their kids, and domestic violence increases significantly. Witnessing violence between parents is deeply traumatizing for children—so much so that if a person assaults their partner in front of a child, that is its own felony in many places.

Over a half million children are abused or neglected in the US every year, and warmer temperatures sometimes play a role. In 2019, a group of Oklahoma City physicians used a decade of data from their ER to show that child abuse visits track upward with temperature. And an analysis of child protective services records from the lower forty-eight states found that each 10-degree-Fahrenheit rise in summer heat increased child mistreatment reports by about 5 percent. Babies and toddlers are always most at risk.

No household is an island, of course; families live within a community and country, and at that level, too, climate change is increasing conflict and disrupting children's lives. Across the globe, social unrest, ethnic violence, and civil wars are made more likely by extended periods of extreme weather.

Children's own behavior and development also suffer. Like their parents, kids get irritated when hot. They have more trouble cooperating and being kind, and struggle more with learning. As the average outdoor temperature for a school year rises, standardized test scores fall—a finding with serious implications both for the children themselves and for society.

None of this should really be a surprise. Temperature is integral to the body's functioning, down to the molecules in our cells. Fossil fuels have changed the conditions in which the human body evolved, and everything about our children's lives will be affected.

Babies and Football Players

Although pediatric heat illness is often missed, two groups of children make heat-related headlines every year: high school football players who collapse on the field, and babies who are inadvertently left in their car seats by distracted caregivers.

Few tragedies are as wrenching. Harried parents, juggling multiple children and errands, forget a sleeping infant or toddler in the back seat—and return hours later to horror. On a summer day, an infant left in a closed car can reach a core temperature of 104 degrees within minutes, and quickly succumb. These cases were rare until the 1990s, when an invention meant to save lives, the front seat air bag, led to babies being placed in the back seat and forgotten more often. *Vehicular heatstroke* has killed more than 960 American children since 1998, about 38 per year, but the number jumped to over 50 in 2018 and 2019, likely due in part to rising temperatures. Climate clearly plays a role, as hotter states in the southern half of the country tend to have the most cases.

In light of this growing toll, the federal government passed a law in 2021 that will require new cars to have a "rear seat check" reminder that activates when the engine is shut off. The major car manufacturers have already agreed. Some automakers are going further, installing more sophisticated "detect and alert" systems preferred by safety experts. Smartphone apps and car seat sensors are also available for parents with older cars. Hopefully, these new technologies can continue the trend seen in 2020–2021, when fatalities fell significantly due to COVID-19 restrictions. Yet as hot days increase, so will opportunities for a moment of inattention to lead to heartbreak.

Teen football players also regularly make heat-related news. Over 9,200

American high school athletes are sickened annually by heat; their ER visits for heat illness have more than doubled since the late 1990s. But young football players are most likely to die from heatstroke. Their fatal cases tripled from 1994 to 2009.

The climate crisis is largely to blame for these trends. Athletes are training and competing in hotter temperatures, of course. But because warmer air holds more moisture, they must also cope with rising humidity, which reduces the evaporation of sweat and thus their ability to cool. Football players are more at risk because their uniforms and padding cover and insulate the skin. Many players who die from heatstroke are also obese (86 percent are linebackers), which impairs heat regulation. Global warming has thus given the childhood obesity epidemic another layer of urgency.

Like Joey and Cody, these teen athletes are stricken with exertional heatstroke, and it could be prevented. Coaches need to encourage frequent hydration and make clear that refusing water is not a sign of toughness. They should assess field conditions with a small "wet-bulb globe temperature" device that measures temperature, humidity, wind speed, and solar intensity. This is especially important in the first two weeks of practice when players have not acclimated, and most heat deaths occur. Experts also recommend "cooling tubs" on the sidelines to immerse overheated teens in ice and water, an intervention that is already saving lives.

Perhaps if ice-stocked freezers and a cooling tub had been installed in a trailhead bathroom, Cody's case would have unfolded differently. The sooner kids are submerged in icy water, the better their chances of survival; every minute makes a difference. But I went to the Apache Wash trailhead, and looked. There wasn't even running water.

Valley of the Sun

The paramedics ran hard when they brought Cody down. The news crew from ABC15 caught it from overhead; everyone in Phoenix could see it.

The men encircled the gurney in dark blue fire department T-shirts, bouncing as they descended the trail. One of them jogged by Cody's head, pumping the ventilation bag. As they neared the parking lot, more men streamed up from the ambulances and fire trucks to meet them; the boy looked as if he were floating on a cloud of blue. When they reached the medical helicopter, most of the men peeled off and just four of them ducked under the blades. One of their caps blew off, and swirled away, as they loaded Cody for the trip to Phoenix Children's Hospital.

Brian Flom had received a call from his ex-wife, and was trying to understand what was happening. *Don't go to the trail*, she said, *go directly to the hospital*. That day was an end and a beginning. It cleaved his life in two.

The man who took Cody Flom hiking surely knew the risks. It would have been hard for anyone to live in Phoenix and not know. For weeks, the nightly news had been filled with footage of heat rescue and death—of helicopters touching down at trailheads, of emergency medical teams running alongside gurneys. Or worse: not running, because time no longer mattered.

The victims that summer included recent migrants and tourists, but also longtime residents. They ranged from a nineteen-year-old who had moved from Seattle to start a new life, to a fifty-seven-year-old German researcher who was in Arizona attending a conference. They included the very fit—a woman who was a personal trainer and young men who played sports—and a middle-aged woman just out for a city walk.

Other victims did not draw the cameras' attention, because they never left their homes. Mostly elderly and alone, they died—as happens each year in Maricopa County—from air conditioners failing or costing too much to run; from having no one to take them to safety. Some of the victims were homeless. They were the kind of people Eric Klinenberg meant when he noted sadly, in his book about the 1995 Chicago heat disaster that claimed 739 lives, that heat waves get less attention than other natural disasters because they are "silent and invisible killers of silent and invisible people."

Everyone in Phoenix knows heat. Yet an invisible enemy is easily underestimated, often by those who know it best. They don't see it morphing, degree by degree, into something new. "I grew up here," Joey's mother said. "It's just part of life. I didn't think of it as deadly." Phoenix has always been hot, locals shrug, and that is true.

But heat risks have grown dramatically since the city's first settlers set up their camps. I asked Nancy Selover, the state climatologist, to pull archived weather records so we could compare the climate of Robert Colley's time to that of Cody Flom. The results were startling. In the century between Robert's 1919 arrival and Cody's 2016 hike, Phoenix's average annual number of 100-degree days increased 52 percent. While Robert typically saw 4 days a year over 110 degrees, Cody saw 21. The period between the first and last 100-degree day of the season lengthened by more than a month; the average annual temperature rose by over 6 degrees. And though television reporters did not call it to the public's attention, for 22 days in July 2016—the month of Cody's hike—both daytime and nighttime temperatures in Phoenix exceeded historical averages.

We face a different challenge than the healthseekers did, more ominous than the tuberculosis and influenza they fled, and harder to escape.

Phoenix's leaders clearly understand what is happening. Multiple agencies and groups are pushing for the city to cool via adaptation: planting trees, increasing shade and solar panels over parking lots and buildings, giving out bottled water, painting roads light gray, and using new construction materials that absorb less heat. Air-conditioned "cooling stations" are scattered through low-income areas, where heat mortality is 20 times higher than the rest of the county. Sadly, their use was curtailed by COVID-19, creating a double threat to these communities during the pandemic.

Dave Hondula, a heat-health researcher at Arizona State University, is optimistic about the city's ability to survive. He told me that Phoenix "may be uniquely positioned to serve as a positive example for other cities for not only how to cope with heat but also how to thrive in a future that will be warmer than the present." He pointed out that, despite rising temperatures, heat mortality in the twentieth century declined dramatically because of air-conditioning, proving our ability to adapt.

We didn't discuss the obvious footnote: that as air-conditioning cools our buildings, it heats the planet; as it keeps children safe today, it endangers their tomorrows. That's because for now, most air-conditioning is powered by fossil-fuel-burning electric utility plants, one of the country's biggest sources of carbon pollution. And the refrigerants that leak from air conditioners, hydrofluorocarbons, are among the most potent greenhouse gases known.

Air-conditioning is ubiquitous in Phoenix, of course. In the coming decades, as temperatures rise, its use will skyrocket around the world. Thankfully, scientists have developed new, climate-friendly coolants to replace hydrofluorocarbons, and we now have an alternative to air-conditioning and furnaces, the heat pump, which can also use these safer refrigerants. But if we don't rapidly adopt this better technology and power the grid with sustainable electricity sources like solar and wind, cooling our buildings will worsen the very problem we're trying to address.

The voters of Arizona should be told this. Yet the year after I traced Cody's steps on the trail, they rejected a plan to get half the state's electricity from renewable sources by 2030. Local utilities had spent tens of millions convincing them that a shift away from fossil fuels was not in their best interest.

In the meantime, Phoenix continues to grow, and get hotter.

The city can do a lot to adapt, and it must. But what will happen if, in the broader world, greenhouse gas emissions aren't curtailed? What is the limit of Phoenix's endurance? 140 degrees? 150? Magical thinking won't prevent those temperatures from reaching Arizona if we fail to change our habits; the laws of physics dictate that they will.

That reality led Phoenix in 2020 to join major cities around the world pledging to reduce their own carbon emissions—because adaptation to the climate crisis is essential, but not enough. It must be coupled with sustainability: ceasing our pollution of the atmosphere, which every day makes the crisis worse.

Phoenix is named for a mythical bird, reborn from the ashes of its prior self. Perhaps it will survive, and thrive, in a hotter world. Perhaps it will be a model for the rest of us to follow, as our own towns and cities heat up; perhaps the lessons learned there will help.

But people will likely venture outdoors and get into trouble, as they do now, all the time. Every year hundreds of heat-sick hikers are rescued from Phoenix's mountain trails. Adaptations to protect them are urgent, but complicated.

The parks department has tried education. Cody's mother's boyfriend walked him past a large flame-colored sign at the trailhead, warning of the risk between April and October. *Each year hikers suffer serious illness or death from heat exhaustion*, it says. The car radio may have reminded him as he drove Cody there: respect the heat, carry enough water, know the signs of heat illness. Don't put yourselves, and rescue personnel, at risk.

Just three weeks before that drive, the city had held a hearing on another idea: closing trails on extremely hot days. Opponents packed the meeting. Hikers who need rescue are victims of their own stupidity, they argued. The rest of us shouldn't be told what to do.

In the face of the uproar, the city decided to ban dogs from trails if temperatures pass 100 degrees; the sign at the Apache Wash trailhead said so. But it voted down limits on people, including any restrictions on children. Taking a twelve-year-old boy hiking up a desert mountain, in the middle of a heat wave, would not be illegal.

Cody's Voice

When Brian arrived at Phoenix Children's Hospital, Cody was in the emergency room. He had arrived at 5:26 p.m. The panic and chaos in the scene come through in the notes of the physician who tried to save him. Cody was found down in the field, the doctor wrote; paramedics thought he had been lying on the trail for over an hour. Later that night, news reporters would stand in front of the hospital and explain that the boyfriend said his phone didn't work, and he'd had to run down the mountain for help.

Everyone who loved Cody is haunted by that detail: that he was left alone.

He was found in a coma; he would never know the men who ran up the trail, in the searing heat, and brought him down. Those men, too, are haunted by what they saw—and have carried it with them. "His case holds a very special place in my heart," one of them told me.

Cody had no medical history, the admitting note states. His lungs were clear; his heart, initially, had a regular rhythm. But his skin was badly burned and bruised, with scattered lacerations. His temperature on arrival was 107.8 degrees. He was packed in more ice, and more cool fluid was pumped into him. Yet his heart soon started to fail.

They had walked a mile out and were on their way back to the parking lot, the boyfriend would tell investigators, when Cody collapsed at the old mesquite.

In the weeks that followed, dozens of people walked up the trail and tied hot-pink ribbons on the tree to remember him.

Brian and his wife, Heather, met me for coffee at a Chili's in north Phoenix the following spring. They have built a legacy from their grief: an organization, called Cody's Voice, to speak about that day and educate other parents about heat's risks. Its name has special meaning to Cody's family. "He had no voice in what happened to him," Brian said.

On hot days, they give out bottled water at trailheads with a photo of Cody, smiling, on the label. Brian has appeared on local television multiple times to warn parents not to take heat lightly. Supporters are given Cody's Voice bracelets, also hot pink. Cody's Boy Scout troop joined in with projects to educate the public about heat and health, and spread his story.

The family has been devastated, no one more so than Brian. Cody was his only child. "It's hard to live with the anger; it changes you," he tells me. His son's life was treated recklessly; it is a pain that doesn't end. "It was completely preventable," he says.

☼

Marguerite and Albert Colley came to Arizona hoping that their son could escape some of the biggest calamities ever to strike humanity. The White Plague and influenza pandemic were overwhelming, terrifying events—like two COVID-19–type disasters at the same time, in an era when medicine's understanding of infectious diseases was in its infancy.

Yet the Colleys did not despair. They acted to protect their child. I wondered, as I studied their faces in the old photographs: what would they have given to look at Robert, playing in the yard of the Sunnyslope school, and know that their efforts had saved him?

The parents I know today face a new type of catastrophe.

With each successive year since Cody Flom walked up the mountain, heat deaths in Maricopa County have broken records, and many more hearts. At least 579 people were killed by heat in 2023, almost *quadruple* the toll of 2016.

By late 2021 the Phoenix parks department decided it had finally had enough, and banned hiking on its trails in extreme heat. This time, objections were few. A dozen firefighters had developed heat illness during the prior summer's rescues, and their union had warned that paramedics' own lives were being endangered by rising temperatures.

The predictions I had read that first night in my Phoenix hotel were coming true faster than I'd expected. But not just in Arizona. A stunning spike in global temperatures would make 2023 the hottest year in human history, annihilating the record set in 2016. Phoenix's summer would be its hottest ever, with fifty-four days crossing 110 degrees—more than twice the number seen the year of Cody's hike.

Another hotter year will soon come.

Heatstroke is treated with extreme urgency; minutes make the difference between life and death. Joey Azuela is alive because he was cooled so quickly. Yet as the world watches temperatures climb, we drift, and delay; we risk pushing the planet to tipping points of rapid and uncontrollable changes, from which we cannot recover. The speed of our response is everything. It will determine not just the type of future our children have, but whether they have a future, at all.

When I departed Phoenix, the flight crew asked us to leave the plane's window shades down while we were on the ground, to keep its interior from overheating. After we were aloft, I raised the shade and looked down at gleaming buildings and glaring desert, at a river of cars on the freeway. Thousands of sun-soaked rooftops watched us go. Though Arizona is one of the sunniest places on Earth, fossil fuels still provide the biggest share of electricity in the state. The plane, I knew, was spewing exhaust, and the runway, roads, and buildings were baking in the heat, which they would radiate till long after dark, preventing the city from cooling.

We all go home to our children at night, and tell ourselves they are everything. But it is so easy to look across the landscape of the Earth and not see them. Over one million children live in and around Phoenix; almost two billion children in the world. They can't control what is happening; the youngest don't even know. They trust us to keep them safe, and we are failing them.

Chapter Three

HOME

"Mommy," he said, "I was scared for a *trillion* minutes."

His small hands were on her cheeks, locking her gaze like a magnet. He was sitting on her lap, in the spare bedroom of the friends who had taken them in. *You were so brave*, she had said a moment before, and brushed his bangs aside to kiss him on the forehead. She repeated this ritual nearly every day. But this time he had gripped her face and wanted her to listen: he did not feel brave.

His serious expression usually made her smile. That day it froze her in guilt. She didn't feel brave, either; she felt like she was drowning.

Her son had seemed fine for several months, she thought. Now images of that night were rushing in when he slept, as uninvited as the flood. He saw water fill the house again. It lapped under his door, reached for his mattress like a stranger, raised him off his bed. The nightmare would end when the flood rose over his head and he realized his parents had forgotten him.

He had heard her shout: *Grab the twins!* Why hadn't she said, grab Lucas?

"He doesn't understand," she tells me through tears, "how fast it all happened."

He had gone to bed that night safe, beneath the oak branches that sheltered the houses around him; beside trunks that held his world in place. He woke up somewhere else. His parents were shouting. He thought he heard splashing, like in the bathtub. When he climbed out of bed in the dark to see what was happening, he slipped and fell into water.

Mommy ran in when she heard him cry. She was trying not to seem scared. As she lifted him he saw his toys drifting around her legs. He looked back as they waded out of the room. His bed was surrounded by a black lake.

Sometimes still, he remembers leaving his Spider-Man floating in the water, and feels bad.

They went to the living room. Water was flowing in like a river: beneath doors, through windows, and from the hallway closet, where a wall had torn away. In his half sleep, he thought maybe they were on a boat, but he had forgotten how they got on it, and now the boat was sinking. His mom couldn't set him on the floor, so she put him next to a pile of photo albums on the couch, which was up on cinder blocks. Soon the couch started to float; the ceiling got closer and closer. For a while afterward, whenever he was inside a building he was afraid to look up.

His parents were upset. They took turns dialing their cell phones, but nobody could come to help. The water was now up to their waists; it was going to swallow them. They kept looking out the window. Daddy said: *We have to go now, or it'll be too late.*

They all went out the front door. It was dark but he could see their car in the driveway, already filled to the top of the seats. He felt even more scared.

Lucas was hugging his mom tight; the twins—his brother and sister, only one year old—were with his dad, one in each arm. Max, the dog, was swimming alongside. The rain was hitting the water so loud, it was like drums banging in his ears. They could hardly hear each other. *We can't get up on the roof with these babies*, his mom yelled. *What if one of them falls*? They began wading across the cul-de-sac to their neighbors, who had a two-story home. But the water was already too deep; they couldn't make it.

Just then they heard a shout. The neighbors' teenage sons were inching toward them on a small rise at the end of the cul-de-sac, along the fence that ran between their houses.

One of the boys took Lucas and put him on his shoulders. Lucas wrapped his arms around the boy's forehead. The other boy grabbed Max's collar as he swam alongside. Mommy and Daddy each took one of the twins. The teens and grown-ups held on to the fence and crept forward, fighting the rushing water that tried to pull them away.

Mommy told Lucas to look ahead. But he couldn't help it, he kept looking over his shoulder at their house, getting smaller behind them as the water rose over its windows. He knew the street where he rode his scooter was only a few feet away, but he was a different boy now, in a different place.

When they got to the neighbors' house, the flood was rising there, too. Lucas wondered how high it could go. Inside, Max climbed their stairs a few steps to shake himself dry, then turned to bark and whine at the water still pouring into the first floor. He kept looking at Lucas like he was supposed to fix it.

Daddy didn't stay. He said the fence was backing up the flood like a dam, making it higher near their house, and he would have to pull it out. Daddy is really tall but when he splashed away from them, he was covered up to his chest.

They all went upstairs with the neighbors. Mommy opened a window so she could hear Daddy. Lucas's pajamas were wet and cold. The twins were crying; Mommy told them *shh shh* they were all right. She kept her eyes on Daddy, but also on their house across the street, disappearing beneath the water. She was crying, too.

The sun was coming up. Mommy put her arm around Lucas. He was doing so great, she said. She was so proud of him. *We're going to be okay, okay?* They watched Daddy heave a section of fence against the ugly lake that was taking over the world. Daddy was trying to save them.

Something was drifting by the window, on the surface of the water. It was a cluster of fire ants, whose bite every kid in Texas feared. They were clinging to a tiny branch, and to one another. Lucas didn't feel afraid of them just now, only sorry. He sat, quiet and shivering, and watched them pass: a wriggling mass of life, helpless against the current.

Turning Up the Dial

A few months later, I stood in front of the home that Lucas and his family had fled. His mother, Tess, had given me the address and arranged to meet me there. But at the last minute, it was too much for her. She was still struggling with what had happened that night.

Their house was in Meyerland, a modest, tree-lined Houston neighborhood that had been hit especially hard—which, in Hurricane Harvey, was really saying something. Before leaving for Texas, I had watched a video shot here by a local reporter and his cameraman the night after the storm started. They had gone down these streets in a boat, scanning with a searchlight; mile after mile of ghostly, half-underwater houses emerged from the darkness as they passed.

Now, on a sunny morning in early 2018, the water had departed but its damage was everywhere. Mounds of silt sloped against walls and wrapped swirling tentacles beneath broken windows. Garage doors looked as if they had been grabbed at both ends and twisted violently, exposing open triangles at each corner. I peeked through one and saw a doll's arm, tinged with black mold, sticking out from a pile of leaves that had collected inside.

At the one-story contemporary ranch house where Lucas's family had lived, weeds consumed the yard. A small security sign, on a wire stuck in the mud, warned *This Property Is Protected.* The floodwaters, as if taunted, had bent the wire nearly to the ground, then torn open the front wall of their house like a zipper, from floor to roof.

Crossing the upper end of this breach was a smudgy horizontal line: the high-water mark. I crouched down. What had six feet of water looked like, to someone only three and a half feet tall?

Lucas, just five years old, had met a deeper, wider, and angrier flood than any current American adult would remember. "As Harvey's rains unfolded," the *Washington Post* reported, "the intensity and scope of the disaster were so enormous that weather forecasters, first responders, the victims, everyone really, couldn't believe their eyes. . . . This flood event is on an entirely different scale than what we've seen before in the United States."

I had watched it from a distance: an immense one-eyed monster that parked itself over Houston and refused to leave for days. As the hurricane spun in place, lifting water from the Gulf of Mexico and dumping it on land, it would produce the continent's most intense rainfall ever: over sixty inches in a single storm. "I can't even describe the sound," one mother told me, calling from the hotel room where she and her children had been living since. "It was like the rain was trying to break through the roof."

More than 60,000 people in Harris County had to be rescued. Nearly half of all housing in the Houston area was damaged, affecting hundreds of thousands of children. Lucas's story, in other words, was not unique—and more kids from across the country would soon join him. That year, 2017, began with news that 2016 had been the hottest year ever recorded on Earth; it would end as the most destructive year of natural disasters in the history of the United States.

Most notable: a trio of record-setting hurricanes—first Harvey, then Irma and Maria—that pummeled the Gulf states and Caribbean in a rapid-fire, four-week period. As those communities staggered, unprecedented urban wildfires struck Northern and Southern California, reducing entire neighborhoods to ash in hours.

That fall was a turning point in Americans' experience of the climate crisis. Millions were left shell-shocked and reeling; hundreds of communities were damaged by wind, water, and fire. But for all the attention these events received, their physical and emotional toll on children—smaller, unable to flee without help, and lacking the vocabulary and maturity to process trauma—were mostly unseen by those who tallied their costs.

What was destroyed, for most people, was the place that they called home. And what was lost, for many children and their parents, was all that home should mean: security and stability, the anchor of a family's past.

Natural disasters have always plagued us; the events themselves are nothing new. But a warming world is turning up their dial, and with it, the potential for trauma. Though some years are better than others, weather-related catastrophes are clearly trending worse over time: becoming more frequent, more powerful, and more destructive. Globally, natural disasters have increased fivefold over the last half century. Extreme weather events—the worst examples of these disasters, like 100-year floods and Category 4 hurricanes—are growing steadily more severe, and more common.

Yet statistics about catastrophe tell so little about being in one, especially for the youngest among us.

For some children, the destruction of a house, a school, and the fabric of community is a trauma that can alter mind, body, and future. Home is where parents and family, teachers and friends are; where children learn to love and be loved. It is where they write their story about the world: whether it is trustworthy and good, or hostile and frightening.

Describing worsening natural disasters with facts about their strength and cost overlooks this lasting emotional toll. A home is more than a building; a little boy's fear is not a number. For Lucas and many children like him, the damage of a hurricane or wildfire doesn't fit on a spreadsheet.

Young Minds in a Chaotic World

"I don't like talking about it because it just reminds me," she says, and starts to cry.

I was leaning forward, trying to hear her. "You don't have to talk about it, Sophia," I say quietly. "Only if you want to." She is a petite fifteen-year-old from an immigrant family. Her black hair is held back by two flowered clips that match her lavender sweater; her shoulders are heaving with sobs. We are in the counselor's office at her school in the Gulfton neighborhood of Houston, where many lower-income families were severely affected by Harvey.

"Since then, every time I walk through the streets, I just can't . . ." Her voice breaks. "I remember it, every second. It's hard."

An image keeps popping into her mind: her parents laying out all the family's shoes on the kitchen table before the storm. They were trying to keep them dry. "Why does that image haunt you?"

"Because we just . . . we lost . . . *everything*," she says. The shoes and the damaged table later had to be thrown out, along with nearly all that the family owned. For months, she and her parents and brother have been sleeping on the bare floor of their stripped and molding house.

Now the sights and sounds of that week—the flutter of helicopter blades, the patter of rain, a puddle on the sidewalk—bring back awful memories. Her hands shake as she lists these everyday cues. And she still has bad dreams, she says, about the long night when her family huddled in the attic, fearing for their lives.

Sophia is describing symptoms of post-traumatic stress disorder, or PTSD.

Though we often associate PTSD with soldiers who've been through a war, it can occur in anyone who survives or witnesses a terrifying event. Like adults, children are diagnosed with PTSD when they keep reliving a traumatic experience through nightmares and *intrusive thoughts*: troubling images that come to mind unexpectedly, often triggered by some reminder. They may suffer from insomnia, be anxious when separated from parents, and avoid places or topics that remind them of what happened. These children struggle—sometimes briefly, sometimes for years—to put the event behind them and return to their normal state of mental health.

Natural disasters always leave an epidemic of PTSD in their wake. As these catastrophes escalate in the climate crisis, pediatricians and psychologists are understandably concerned about their effect on child survivors—a worry that stems, in part, from the fate of children who lived through one of the deadliest natural disasters in American history.

❁

Hurricane Katrina displaced more than a million people from their homes in 2005, and killed over 1,800. But I can't think about that storm without remembering one person, Nia, a teenager who came into my clinic five years later. She was nine years old when her grandmother, who had been her caregiver, died in Katrina's aftermath. In the years since, she had stayed with an aunt in Houston, but they had just moved to Reno. She was now fourteen.

Her aunt told me privately that Nia had been "obsessed" with memories of her grandmother's death and their ravaged house for quite a while. She was fidgety in the exam room; one of her teachers sent a note saying he suspected ADHD. Nia was tall for her age, and lean, with neat cornrows pulled to the back and falling to her shoulders. She smiled when she greeted me but made eye contact only fleetingly and always paused before responding to my questions—as if there might be a wrong answer to how she felt about school, or whether she played sports.

I told Nia how sorry I was about her grandmother, and asked if she had a picture of her. She hesitated, then reached into her backpack and pulled a photo from her wallet. I saw a fifty-ish woman with a warm smile, hugging a younger Nia. "This is her," she said.

Later that year, I was not surprised when a study of the "children of Katrina" showed that a third of those displaced by the storm met criteria for PTSD, anxiety and depression, or ADHD. Or that rates of "serious emotional disturbance" were nearly 5 times higher than in children not traumatized by the disaster.

Other researchers would find that teens who had lost a home or loved one, as Nia had, were more likely to abuse drugs and alcohol. But a third type of event also strongly predicted emotional distress: the loss of education.

Like Nia, hundreds of thousands of New Orleans children could not return to their schools, either because of damage to the buildings themselves or displacement of families, many of whom relocated multiple times after the storm. The hurricane led to large gaps in education for tens of thousands of children, who as a group tested significantly below their peers and never caught up. In a ten-year follow-up in 2015, *The Atlantic* reported that "Katrina-era adolescents . . . often suffered such high levels of trauma and instability that learning became nearly impossible. It was 'like throwing seeds at cement,'" said one educator.

When I read that quote, I hoped the children it described would never see it. Though Nia's struggles were obvious, her heart and mind had clearly not turned to cement. With medication, therapy, her aunt's love, and her own hard work, she finished high school shortly before that article was published. The last time I saw her, when she aged out of my pediatric practice on her eighteenth birthday, they were moving back to Houston to be closer to extended family. She told me she hoped to work in an office. I hugged her good-bye and wished her well.

A few years later, as I was speaking with Sophia in Houston, it suddenly occurred to me that Nia was living nearby and might have faced something similar. That she was one of roughly 100,000 Katrina refugees who fled to Houston and had now been hit by Harvey, another record-setting hurricane and flood.

I called my office to get her address from a Christmas card they had sent me after their move. The next day I located her aunt's house, surrounded by a chain-link fence covered with dark green tarp: the mark of a Harvey-damaged home. Their neighborhood echoed with the *whoosh-thomp* of torn-out Sheetrock and carpeting being heaved through windows to the ground. Looking at those moldy piles on the curb, I worried about Nia's lungs. She was one of many New Orleans children whose asthma symptoms had spiked after Katrina because of increased mold in their homes.

I searched for a mailbox to leave a note. If there ever had been one, it was washed away.

Climate disaster had chased Nia from one place to another. In the past, her story might have been chalked up to bad luck. But before this century, experiencing two such extreme weather events in her short lifetime would have been nearly impossible. Harvey alone was unprecedented; no prior North American storm had brought so much rain.

We know that its magnitude was due to climate change. Part of Nia's bad luck, then, was simply being born into this generation of children and young adults—who were left, decades ago, to the global crisis we knew was coming.

That negligence led directly to what happened to her. A warmer world is a more chaotic world, driven to extremes of weather and thought, plagued by violence from nature and man. A world where Nia could survive one worst-ever storm and walk into another; where Sophia's family could flee Guatemala because of crop failure and conflict spawned by a historic drought, and then lose their next home to a historic flood.

The repetitive trauma experienced by these girls is one of the most concerning health impacts of climate change, particularly for children whose families have few resources with which to recover from even a single disaster. Yet hurricanes and floods are only one source of such trauma, and the Gulf is only one place it will strike. Shortly after my first trip to Houston, I met another little girl whose life would take a similar turn, in a neighborhood two thousand miles away.

Shani, like Lucas, was just five years old when her family escaped a natural disaster linked to climate change. But in her case the menace was not water, but flame.

I met Shani and her mother and seven-year-old brother at a city park in 2018, not far from the neighborhood they had fled. Her mother told me their story as we sat on a bench and watched Shani and her brother kick a soccer ball. Afterward, with her permission, I drove to see what was left of their house, too.

Coffey Park had been a lovely expanse of middle-class homes in the center of Santa Rosa, California. In photos residents posted online, children ran through fenceless yards shaded by leafy trees. Out Shani's front door there had been a willow taller than her house, dangling a swing where kids congregated after school.

Now, as I looked west and north from their lot, all that remained of Coffey Park was an eerie plain of ash stretching as far as I could see. Block after block of empty streets led to driveways shorn of their houses. The odor of charred wood, which had stung my nose from miles away as I approached the town, was so strong I had to pull my sweater over my face. Here, in the early hours of October 9, 2017, the Tubbs Fire had reduced thousands of buildings to soot in hours, and killed twenty-two people.

Shani had been sleeping peacefully with her stuffed tiger when the loud bang of a door, caught in the howling wind, jolted her awake. She screamed *Mommy!* and her mother came running, hugged Shani tight, and told her it would be okay. They went downstairs. The house was dark and filled with smoke.

Through the living room window came a strange orange glow. The willow tree in the front yard was on fire. It towered over them like a giant torch, flaming limbs blown horizontal by the wind. Her old friend looked like it was pleading for help.

Flashlights were flicking on inside neighbors' windows, barely visible through thick smoke. Pieces of burning debris drifted and tumbled through the air, igniting every yard. They heard *BOOM! . . . BOOM!*, like a monster stomping their way, as the fire reached yet another car or propane tank. The smell of the smoke kept shifting from burnt wood to melting plastic and vaporized chemicals.

Her mother told Shani to stay in the house. Then she went outside, turned on the garden hose, and blasted it at the willow. A stranger drove by and shouted *GET OUT NOW!* over the roar of wind and fire. Her mother froze for a second with the hose in her hand. Then she threw it on the ground, rushed back inside, told Shani to put her shoes on, and ran upstairs to get her brother, still asleep in his bed.

They couldn't find the dog and thought they might have to leave her; she was cowering in the garage. There wasn't time to grab anything. Within minutes they were all in the car. Shani saw her mother's hands shaking as she buckled her into the car seat. Her mom wouldn't realize until later that with the electricity out, an elderly neighbor couldn't open her garage door to escape. She would relive those moments in the driveway with guilt and horror, wishing she could go back to help.

Only a block from their house they got stuck in traffic. All the neighbors were trying to evacuate on the same two-lane street. Shani's mother watched the flames in the rearview mirror, wondering if they should abandon the car and run. The kids peered out the side windows as their friends' houses were engulfed. A fire truck rushed by without its siren, its crew stunned and silent.

They inched out to the main road and eventually made it to Shani's aunt's house, along with other family. But they would soon be on the move again. "It came to a point where we did not even feel safe at my sister's," her mother said, "so we decided to head west, out of town. When we finally got to open countryside, I looked across the valley and saw the line of fire ripping through the place I have always called home. It was the first time I really let myself cry." They spent the night in their car.

A photo of the blackened and bent trunk of their willow tree would appear in newspapers the next day, a symbol of what had been lost. When I asked Shani's mother what she had felt, she said, "Shock, I mean, the power of it . . . one minute everything's here, and the next, it's all gone." Shani said she missed her tiger.

I kept in touch with them as the neighborhood rebuilt. After almost two years of temporary housing, they moved into a new house, on the same lot, in late 2019. The local news celebrated Coffey Park's revival as many families returned. But within weeks the area would evacuate again because of the Kincade Fire. *Eighteen* major wildfires have raged through Wine Country since 2015; Shani and her brother have had to keep a wary eye on smoke over and over again. In the fall of 2020, they looked up at skies turned blood orange by nearby fires—images that made international news as a sign of climate apocalypse.

I grew up nearby, but never saw a single wildfire as a child. I don't know what it's like to have your view of the world shaped by the recurrent, terrifying threat of losing everything. Lucas's situation is similar: the number of heavy downpours in Houston has doubled in the last thirty years, leading to ever more frequent floods. How will these types of experiences, shared in some form now by millions of children, affect long-term mental health? Can parents, pediatricians, and psychologists do anything to help them cope with mounting natural disasters? I had met someone in Houston who was studying those very questions.

The Children of Hurricane Harvey

When Harvey made landfall, Dr. Julie Kaplow was still unpacking her office in a tower of Texas Children's Hospital.

Kaplow is a child psychologist and an expert in children's reactions to loss, mistreatment, and terrible events. She had been hired by the hospital and its associated medical school, Baylor College of Medicine, to establish a pediatric trauma and grief center. Her timing, sadly, was perfect.

Only two weeks after she started the new job, before her books were even out of their boxes, the most destructive flood Americans had ever seen would devastate her city. Kaplow would soon find herself starting a project she hadn't anticipated: the Harvey Resiliency and Recovery Program, for children traumatized by the storm.

When I met with Dr. Kaplow at her clinic in early 2018, an estimated 22,000 kids in the area were still homeless, mostly living in hotels. The households of over a half million children had been damaged or destroyed. At that point, she told me, her clinic's Harvey program was getting roughly one hundred referrals a month from pediatricians, teachers, social workers, and parents: anyone who knew a child needing help.

Kaplow understood that a storm of Harvey's size could create a wave of suffering for children in southeastern Texas that, if not addressed, might ripple through their lives. But she also knew that not every child affected faced the same risk. Many children would experience only short-term emotional distress and recover, while others might be forever derailed from their full potential in school, relationships, and work. What made the difference?

The answer has been pieced together from studies of children who survived past major disasters, including hurricanes Katrina, Andrew, and Sandy, and the massive Fort McMurray wildfire in Canada. Between 30 to 70 percent of the children who were impacted directly by these catastrophes displayed mental health and learning problems that persisted for more than a year. Their stories have taught us that the odds of long-term symptoms depend on events before, during, and after the crisis. And that certain experiences are more powerful than others.

Using these findings, Kaplow and her colleagues began community screening of children affected by Harvey, hoping to identify those in need of support. While most prior researchers relied on parents' reports, Kaplow's team developed a simple questionnaire for the children themselves, specifically those eight years and older who could describe their memories and feelings. In 2019, the year after I met with her, they published their initial results, and confirmed much of what I saw in the kids I interviewed.

Their analysis showed that two experiences during the hurricane could, by themselves, strongly predict whether a child would develop both post-traumatic and depressive symptoms. First was thinking that "My family and I might get badly hurt or die." This was not surprising, since trauma's emotional impact is often rooted in the threat or reality of physical harm.

Second was a child's report that "My pet got badly hurt or died." Fewer than a hundred people died in Harvey, but almost 7 percent of the children in Kaplow's program reported the injury or loss of a pet, and many were understandably heartbroken.

Yet a dozen other events during the hurricane could predict either PTSD or depression or, when put together, add up to predict these symptoms. Being temporarily separated from a parent, for example, or just leaving a pet behind. Getting hurt, witnessing someone else being hurt or killed, needing to be rescued by boat or helicopter, or having a parent who was an undocumented immigrant and afraid to call for help. These circumstances were dose-dependent: the more of them a child experienced, the greater the odds of PTSD and depression.

The period after the flood was also important. Many children returned to destroyed homes, cars, schools, and belongings. Some stayed in shelters or moved multiple times. Children whose families struggled, post-Harvey, to meet basic needs—housing, food, clothing, and transportation—were far more likely to have emotional difficulties, underscoring that poverty always exposes young minds to greater risk.

The biggest predictor of mental health problems, though, was often not what happened to a child *during* and *after* Harvey, but what had happened *before.* Children who come into a disaster with a history of trauma—who have been abused or neglected, for example, or survived prior life-threatening events—are much more likely to be impacted emotionally, and for a longer time, than children without these histories. Children who are bereaved, like Nia after she lost her grandmother in Katrina, are especially vulnerable in a subsequent natural disaster.

"Most kids are resilient," Kaplow told me. "In the immediate aftermath, they might have nightmares and come into your bed at night because they're scared. Or they might relive what happened for a while. That's to be expected and will usually dissipate. But if we look carefully at a child's history and find prior major adversities, that's a red flag for us that he or she is more at risk of long-term distress."

Past misfortune, in other words, tends to worsen a child's response to current misfortune. This is a daunting problem as natural disasters rise in number and force, and children are exposed to them again and again. It means that each time a child experiences the terror and loss so common in these events, the odds of long-term emotional and intellectual difficulties go up. Among the distressed children referred to Kaplow's program, 1 in 5 had already lived through a hurricane, flood, or tornado when Harvey struck.

These and other, quieter disasters, like heat waves and droughts, also tend to unleash chaos in children's homes. Adults under severe stress, especially those who have poor coping skills to begin with, are more likely to mistreat each other and their children. Those private traumas, too, inflicted by caregivers who should be giving comfort, become part of a child's history.

They also show that children's mental health cannot be viewed apart from that of their mothers and fathers. Kids are very sensitive to their parents' feelings—to facial expressions, tone of voice, and other cues—and their reactions to disaster depend largely on how parents respond. "You can do whatever you want in therapy," Kaplow observed, "but that won't help the child whose parent is really struggling." Like a flight crew instructing us to put on our own oxygen masks first, Kaplow emphasized that parents need emotional support in order to be present for their children. If they are severely distressed and struggling to function—as 25 percent of adults impacted by Harvey self-reported in the two years after the flood—they may need therapy.

I completely agreed, but both of us knew: easier said than done. Affordable mental health services for adults and children are always in short supply, but never more so than after a disaster. The hospital's Harvey program filled part of this gap by offering short-term individual therapy to kids whose questionnaires revealed significant storm-related distress. The therapy focused initially on teaching children basic coping skills: how to identify their feelings, regulate emotions through deep breathing and other exercises, and change unhelpful thoughts. (For example, if a child expressed that "Houston is a dangerous place," he might be encouraged to say to himself instead, "I'm safe now and there are adults who are here to protect me.")

Early results showed that these efforts were working. The program reduced rates of PTSD and depression in storm-affected children, as well as the frequency of *maladaptive grief*—grieving that keeps kids emotionally stuck and unable to move on, or that makes it hard for them to carry out daily activities. It also improved the kids' school behavior and peer relationships.

The need for these services only grew after I met with Dr. Kaplow. By Harvey's two-year anniversary in 2019, the number of children being referred had not only failed to drop, it had more than doubled, to 250 a month. Clearly, for some hard-hit families, the storm continued long after the flood.

And clearly those thousands of referrals are a sobering marker of the broader mental toll of the climate crisis. Harvey was just one disaster out of many; twenty-two weather-related billion-dollar events struck the US in 2020 alone, a new record. When compounded by the burdens of the COVID-19 pandemic, these calamities left countless children struggling. The trend shows no sign of slowing: 2023 broke the record again, with twenty-five billion-dollar weather disasters that stretched from the tragic wildfire in Maui to historic flooding in Vermont.

Fortunately, the Harvey program's approach can be applied to other catastrophes, because the Trauma and Grief Center is an affiliated site of the National Child Traumatic Stress Network—an alliance of hospitals, universities, and community groups that share resources shown to help traumatized families. These include trauma- and grief-informed assessment tools, preventive interventions like Psychological First Aid (a phone app used to identify individual needs after a disaster), and the Trinka and Sam book series for young children who've survived a hurricane or wildfire. The books are based on the idea that when we show kids how to prepare for and respond to a disaster, and teach them language to describe what they've been through, it is empowering, and reduces traumatic symptoms. Like the Harvey program, they try to give children a sense of agency over their own fate.

I've tried to remember the importance of agency, since my trip to Houston, whenever one of my patients mentions the climate crisis. Children as young as seven have told me they feel sad about what's happening to the Earth. I tell them that many people, young and old, are fighting for the future. I suggest steps they can take in their own home and school to reduce pollution and push for bigger change. Then I remind myself that the battle isn't over. And I rage silently, as I finish my charts, at those responsible, who think that any amount of money is worth what they've done.

Rescue from the Roof

Lucas heard the motorboat before he saw it. It tilted hard to one side as it turned the corner into the cul-de-sac. The three men sitting on its benches wore helmets and bright yellow jackets.

Over here! his mom yelled. Daddy turned from the fence, in the neck-deep water, and started waving his arms over his head.

The men pulled up to the roof like it was a dock. One of them climbed to the window and handed bulky orange life vests to his mom, and she put them on Lucas and herself. *What about the twins?* she asked. Sorry, he said, none of their life vests were small enough for babies. Mommy stared at the man for a second, then climbed out the window, onto the roof. She sat down and reached back for Lucas.

He looked at her outstretched arms. Behind her, across the water, was the top of his house. *Why do we have to go?* His neighbor bent down and said there was no food, or anything to drink, or diapers. Your mom has to get you somewhere safer, she said.

The roof felt rough through his pajamas and on his palms. He held Mommy's hand and worked his way down slowly, so he wouldn't slip, until one of the men reached up and lifted him into the boat. Then Mommy passed the crying babies, one by one, to these men he had never seen before. Last, she tried to coax their dog, Max, out.

No, the men said, Max would have to stay. They needed room for another person trapped in an attic nearby. Lucas started to cry. Max would be scared. He might try to swim after them and get lost.

Daddy yelled, *I'll find another way out. Get them out of here.*

The men were in a hurry, so Mommy grabbed Max and scooted down with him to the boat. They left Daddy behind, in the deep water, alone. Lucas turned around and watched him standing by the fence, until the boat turned out of the cul-de-sac.

The boat's outboard motor echoed off the neighbors' houses as they went down the street, now a strange bayou of floating roofs. The boat's wake splashed against windows. They saw other neighbors on rooftops. The men said, *We'll be back, hang on.*

Mommy held the twins tight, but they were quiet now, and so was Max. Everyone just stared at the brown water, block after block. When they passed the elementary school, Lucas saw its playground equipment sticking up like a wrecked boat. He remembered the story about Noah's flood, which God had sent to punish bad people. Maybe I did something wrong, he thought.

Children's Brains Are Different

When children live through difficult or tragic events, parents usually worry about the long-term effects. Even without research like Dr. Kaplow's, most adults understand—either from our own memories or from the stories of others—that childhood trauma carries special weight in a person's life. But why is this so?

The answer will sound familiar: children's smaller, still-forming bodies are more affected by the environment around them. Just as their lungs are shaped by the quality of air, and their skin, circulation, and nerves are constantly reacting to outside temperatures, brain development, too, is intricately tied to their surroundings; it occurs through interactions with the world.

This wasn't always obvious. For millennia, philosophers and physicians debated whether a child's mind was shaped more by "nature or nurture"—by inherited traits or by learning. Now we know: it's both. Much as a house is built according to a blueprint but the quality of its construction can vary, children's genes determine the structure of the brain, but their experiences shape *how well* it develops. This phenomenon, known as *neuroplasticity*, is most pronounced in babies and young children, and decreases markedly with age. Which is why children's minds are more vulnerable to the trauma unleashed by natural disasters, and why childhood events have long-lasting effects.

Lucas's story can help illustrate this. To understand how his brain responded to the events unleashed by Harvey, though, we have to first look at how it was developing before the storm.

When Lucas was born, his brain already contained most of the billions of nerve cells, or *neurons*, he would ever have. They were floating like stars in the brain's thin outer layer, the *cortex*—also called "gray matter" because of its color. These lonely cells soon began extending tendrils toward one another and communicating, like voices on a telephone line, through chemical signals—connections that are known as *synapses*.

Lucas's synapses stored what he was learning; they are where his memories live. Because he was absorbing so much new information as an infant, they proliferated rapidly; by age two, trillions of them had formed. Had we been able to see a cross-section of his cortex at that point, we would have found what looked like a dense network of wires.

Simultaneously, Lucas's brain started to morph in response to the world in which he found himself. Synapses that held useful skills or knowledge became thicker and stronger as they fired repeatedly, while those that were not used withered and disappeared. This is why an enriching environment—filled with language, love, and learning—is so important in early life. It literally builds the brain.

As this refinement continued, different parts of Lucas's cortex took on various jobs. Areas that control speech, for example, matured on the left side near his ear, while much of his personality housed itself behind his forehead in an area called the *right orbitofrontal cortex*, or OFC. His family, unknowingly, was watching these changes in real time. Each of the developmental milestones he met as a baby and toddler—such as taking his first step or saying his first word—was a sign that another area of cortex had come online.

But not just the cortex. Beneath it, in the white matter that makes up the bulk of Lucas's brain, larger nerves were connecting these different areas of the cortex like cables on a switchboard. They were also linking the cortex to the brain's deepest structures; for example, the right OFC communicated with the *limbic system*, the emotionally important area in the center of the brain, via a band of nerves called the *uncinate fasciculus*. The strength of these white matter tracts also, it turns out, depends on the quality of a child's environment.

We know this because new types of magnetic resonance images, or MRIs, have enabled neuroscientists to see the impact of experiences, good and bad, on children's brains—specifically, on the thickness of the cortex and white matter pathways. They have allowed us to explain psychologists' observations with biologic fact.

Let's take, for example, Dr. Kaplow's statement that a child's emotional response to a disaster cannot be separated from that of his parents.

Psychologists and pediatricians have known for decades that reciprocal love and communication between an infant and parent—the *attachment* we all learn about in parenting classes or Psych 101—is vital for normal development. Babies who are *securely attached*—who have formed a strong, trusting bond with a devoted caregiver—are more likely to thrive. But babies who are *insecurely attached*—who can't trust their parents to provide comfort because they are abusive, neglectful, troubled, or absent—are more likely to struggle with learning, behavior, and empathy.

Brain imaging now shows us why. In the first two to three years, those right-sided brain structures that are so important for personality and emotion actually form through interactions with the parent's mature brain. Like a ship following a beacon, this part of the baby's brain needs cues from the parent's brain to steer itself onto the correct path. Children who are securely attached, as Lucas appeared to be, have much stronger right-brain connections; they literally internalize their parents' love. Insecurely attached and traumatized children have been found to have abnormalities in these structures.

The science of attachment tells us that when Hurricane Harvey hit Lucas's home, his brain was genetically programmed to look for his mother's response, and his feelings were linked to hers because his brain had been forming, since birth, through interactions with hers. When Tess calmly comforted him at the neighbor's as they waited for rescue, she helped buffer Lucas's mind from what had just happened. But as she struggled afterward, understandably, to cope with her own trauma—over her family nearly being killed, losing everything, and being homeless—it is likely that Lucas felt her emotions viscerally, and was more troubled by his own memories.

Attachment science also shows that Lucas's bond with his mother created a brain that could cope better with adversity. His well-developed right-brain structures gave him a positive sense of identity and helped him relate to other people, to have empathy and seek comfort. These traits increased his resiliency: his ability to return, eventually, to his prior state of mental health.

Not every child is so lucky, of course. Which brings us to another of Dr. Kaplow's observations: children with painful pasts tend to suffer most when a storm or wildfire strikes. These children often exhibit less resiliency after a disaster. To understand their troubles, we have to look at how trauma disrupts normal brain development, and can influence a child's response to stressful events even years later.

Toxic Stress

Consider Sophia, the Harvey survivor whose family had escaped climate catastrophe in Central America when she was a toddler. All children feel stress at times, and a certain amount of it, when experienced with loving parents who can teach good coping skills, is important to becoming a healthy, well-functioning adult. But Sophia was likely exposed in her early life to what pediatricians call *toxic stress*: severe, frequent, or persistent stress created by adverse circumstances. Toxic stress differs from normal stress because its intensity or duration exceeds a child's ability to cope.

Named for its capacity to injure a child's brain and body, toxic stress occurs in families facing calamity: poverty, drug abuse, serious illness, violence, or the aftermath of natural disaster. Children experience toxic stress when they must navigate these terrible events without enough hugs and soothing words from a caring adult. Sometimes this happens because parents, too, are overwhelmed, ailing, or separated from the child; sometimes it's because they themselves are the source of the trauma, in the form of abuse or neglect.

When I spoke with Sophia, I obviously could not know if she had been hurt by toxic stress during her family's migration. But I knew that if she had, the record of it was physically imprinted on her cells and even her genes. And that this imprinting, like a well-worn path, had probably guided her reaction to Harvey.

How does toxic stress leave its mark? First, it can affect the architecture of the developing brain, beginning with the structures of attachment. When a baby looking to her parents for comfort rarely finds it—because, for instance, they are focused on staying alive during a civil war—her right brain's attachment circuit may not form the way it should.

Toxic stress sculpts other parts of the brain, too. Recall that a child's neural connections grow or shrivel based on how much they are used. Brain structures that enable us to respond to threat—to sense danger and react quickly—can become overdeveloped in a traumatized child; she may perceive a threat when none exists. At the same time, parts of the brain that help us think logically, plan carefully, and control ourselves may not develop as well, leading to more impulsive behavior.

These physical changes are sometimes large enough to be visible on a brain MRI. But the second way that toxic stress leaves its mark is more subtle. It can alter the chemistry of a child's body by reprogramming her *stress response system*. Consisting of the brain and pituitary and adrenal glands, this system controls hormones that help us react to perceived threats. And it is also highly plastic—that is, molded by experience—in early life.

Here's how it works. When Sophia sensed danger during her family's ordeal, her brain would have instantly, without any conscious direction from her, sent a signal to the pituitary, a tiny stalk-shaped gland in its center. The pituitary would have then sent a second signal to the adrenal glands atop her kidneys, causing them to release *adrenaline* and *cortisol* into her bloodstream. These hormones would have sped up her pulse and breathing, raised her blood pressure, and tightened her muscles so she could run, making her stronger, quicker, and more focused, thus improving her odds of escape.

Children who survive multiple *adverse childhood experiences*, or ACEs, tend to have abnormal levels of adrenaline and cortisol. That's because childhood trauma changes the genes that regulate these hormones. It does this by turning them "on" or "off" with chemical markers, which are collectively known as the *epigenome*—literally, "above the genes." For example, the gene for a protein that tells the body to slow cortisol production can be turned off with a methyl group, resulting in higher levels of cortisol. In adults who were abused as children this gene is highly methylated; the more severe or prolonged the abuse, the more methylation.

For Sophia, such epigenetic changes could have caused her stress hormones to react more dramatically, and return to baseline more slowly, than those of a child without a trauma history. When Harvey struck and her family nearly drowned, this programming would have increased the physiological and emotional impact of that new trauma. It would have made PTSD more likely and more enduring.

Her mind's ability to form in response to experience had clear advantages. In a dangerous environment, where reacting is sometimes more important than reasoning, a "stress-wired" child may have better odds of survival. But that same wiring can be maladaptive over the long term, creating a perpetually heightened state of "fight or flight" that leads to myriad health problems. In animal studies, chronically elevated cortisol has been shown to disrupt neural development and lead to smaller brains, poor growth, and more behavioral and learning problems. In people, adults with a history of several ACEs not only have higher rates of depression and substance abuse but also medical diagnoses such as coronary artery disease and diabetes. The past really is never past; we carry it within us.

Climate change is almost certainly increasing toxic stress for many children in the storm- and wildfire-pummeled regions of the country. Not just because of the emotional impact of the disasters themselves, but because their force can reverberate through a family for months or years in the form of grief, homelessness, financial strain, substance abuse, domestic violence, or forced migration. These chronic tensions mold children's minds while sapping parents' ability to nurture and protect. In other words, natural disasters are global news, but their impact can be profoundly intimate: they may shape who some children *are*.

I lecture on the impact of childhood trauma at the University of Nevada School of Medicine. Inevitably, worried expressions cross a few of the students' faces. Some of them had traumatic childhoods yet made it to medical school. Can't someone survive a difficult early life and become an empathic, healthy adult? The answer is clearly yes. Saying a problem is more likely because of a troubled past does not mean it is certain. And no child, regardless of age, should ever be considered a "lost cause," even if their early years were horrific. I tell the students that some children are probably more genetically susceptible to toxic stress than others. But what seems most critical to resiliency—the ability to overcome childhood trauma—is a caring relationship with at least one supportive adult. I usually see a few nods of recognition.

Love can, in fact, over time, heal some of the wounds of toxic stress. And with a lot of help—for example, from programs like the one at Texas Children's—traumatized children may grow stronger from their experiences. But as one group of researchers put it, "the brain is not infinitely plastic." When trauma piles upon trauma, scars are less likely to fade.

In sum, children may be physically weaker than adults, but childhood is emotionally powerful: it casts a long shadow of protection or vulnerability. Intensifying natural disasters are increasing adverse childhood experiences and toxic stress, and the effects are not easily undone.

Unnatural Disasters

I thanked Dr. Kaplow that day we met, then made my way through the hallways of Texas Children's Hospital, a mini-Manhattan of gleaming towers that form the country's largest pediatric hospital. A father in a NASA cap came toward me down a windowed passage, holding the hands of his two young children and headed toward the Trauma and Grief Center I'd just left. His little boy's shoes scuffed the carpet as they passed a colorful sign on the wall: the hospital was grateful for a charitable donation from Chevron.

I turned and looked behind me as his family rounded a bend. I wondered if he had noticed the sign, or thought anything of it. Much-needed funding for children's health had come, in part, from an industry that was endangering all children—one that guaranteed even worse hurricanes in his son's and daughter's future.

I knew this was an outsider's perspective in Houston. Everyone I met here had friends or family in the oil business or were in the business themselves; the region had been built and enriched largely on that industry's success. But his NASA cap reminded me that there was another major employer here, one that had saved millions of lives from disaster, and might save billions more.

Hurricane warnings are so familiar to us today that it's easy to forget they are a modern wonder. When today's grandparents were children themselves, the most sophisticated tool available to meteorologists was the weather balloon. That changed dramatically on April 14, 1969, when NASA quietly launched a satellite called the Nimbus 3. For the first time, human beings would be able to forecast the weather from far above Earth—not just experience it, with little advance notice, from below.

The Nimbus 3 was one of the most important public health advances of the twentieth century, ultimately as crucial to the well-being of children as vaccines, antibiotics, and water treatment plants. As the first real weather satellite, this small vessel and its successors enabled improved storm warning systems that would markedly reduce global annual deaths from natural disasters—from over 500,000 people in the 1920s to less than 100,000 people in the 2000s—even as populations, and extreme weather, increased.

Nowhere did this feat mean more than in southeastern Texas, site of what is still the deadliest natural disaster in American history. On September 8, 1900, a massive hurricane had nearly obliterated Galveston, a city built on a long, thin barrier island that grips the Gulf Coast. Though meteorologists in Cuba had tried frantically to warn that a disastrous hurricane was headed Texas's way, the new US weather bureau labeled them alarmists and blocked their telegraphs. Between 6,000 and 12,000 people, including an untold number of children, were killed.

More than a century later, Hurricane Harvey, for all the destruction and trauma it wrought, would not kill thousands. Satellite images had allowed Houston's residents to follow the huge, swirling cyclone as it advanced hungrily toward the city, giving some of them time to evacuate. And satellite data had led the National Weather Service to issue a warning more alarming, even, than the one withheld from Galveston: *This event is unprecedented & all impacts are unknown & beyond anything experienced.*

Yet there was another reason the launch of the Nimbus 3 had enormous public health implications. The new satellite could tell us the types of gases in the atmosphere and their temperatures far more accurately than had been possible before. It also allowed scientists to calculate how much of the sun's energy was reaching and leaving Earth. They could see clearly that as pollutants such as carbon dioxide and methane increased, more of the sun's heat was being trapped. Embedded in NASA's first real weather satellite, in other words, was a view to the future: we would now be able to track our transformation of the skies and the warming of the planet.

What NASA's scientists couldn't foresee was that the world would respond to one type of satellite warning, but not the other. That like Cassandra of Greek myth, they would be cursed with the gift of true prophesies, not believed.

For decades since the launch of the Nimbus 3, the agency has warned about the existential threat of fossil fuels. Its satellites have tracked those fuels' alarming impact on the atmosphere. Yet NASA's Johnson Space Center—the place *Apollo 13* famously radioed with "Houston, we've had a problem"—is surrounded by corporations who have spent billions to suppress that science.

These oil giants would be silent when, months after Harvey, scientists showed that global warming had boosted the storm's rainfall total by almost 40 percent. That air pollution from their products had heated the Earth and made the hurricane into the behemoth it was. That the flood now haunting Lucas and Sophia and so many other local children was an unnatural disaster, born of the oil that runs like blood through Houston's economy.

Texas Children's itself was not spared. Cut off by the flood, the hospital could not be reached for days by many of its patients, including almost three dozen children with kidney disease who needed dialysis. As the toxins in their blood climbed, threatening their lives, these children had to be rescued and brought in by all sorts of heroic means, in many cases involving helicopters and boats.

Chevron was unmoved. Although it has changed its rhetoric about climate change, it continues to push for expanded oil and gas production, and funds an industry group that lobbies against solar, wind, and other sustainable technologies. I can't help but think of its executives like Gollum in *The Lord of the Rings,* clinging to gold while falling into a pit of fire. Any profit made from fossil fuels will be meaningless in the future they create. And no hospital will be able to fix what that means for children, in Houston or anywhere else.

Love and Water

Tess found herself standing on the side of the road—but what road, she wasn't sure. When the boat reached shallow water, the crew had motioned for her to get out, asked for their life vests back, and left. As the sound of the motor faded, her heart began to pound. She had a twin in each arm and Lucas was clinging to her leg. The dog was whining. What now? Images of their house being torn open, of water rushing in, flashed by. She started to see lost things: photos and clothes, their cars, Lucas's asthma medication, pacifiers for her crying babies. Everything was gone. And she had left her husband behind.

They needed food, and a place to sit, and clean diapers. But the street was abandoned and flooded, its businesses shuttered and powerless. She could not carry all of them through the water. Her phone wasn't working.

An SUV came slowly up the street and pulled over. A woman she did not know got out, opened the passenger-side doors, and offered to take them to her home. Tess looked at the car's white, spotless interior and her mud-covered family. The woman said: *Look at me. I'm a mom of three boys. I'm not afraid of dirt.*

She and the kids and dog would stay with the woman's family for a day and a half. They bathed, and ate, and slept in the bedroom of one of her sons, a college basketball star who went out into the still-chaotic city and bought pacifiers for the twins. "They *saved* us," Tess told me, crying. For her, this had been the only good thing about Harvey: regardless of race or politics, everyone helped each other.

She didn't know what had happened to her husband until she reconnected with him by phone the next afternoon. He had ended up in a shelter on the other side of the bayou. They were reunited when the water receded. Friends then opened their arms and offered the family a place to stay.

They were safe. But the storm was not quite over.

One morning I was scrolling through photos of the city underwater when my phone rang. Tess had heard I was looking for children traumatized by Harvey.

She told me about Lucas through tears. "I didn't realize he was so scared," she said, chastising herself. He didn't want to leave his dad behind; now he kept remembering him in the water. Whenever it rained, Lucas would repeatedly check that a flood wasn't rising outside the door. A pothole full of runoff would trigger "Oh, this parking lot is flooding!"

She, too, was struggling. She had recurrent nightmares about saving her kids from the flood, or of drowning. Every day she remembered something she wanted from the house and had to remind herself, with new grief, that it was gone.

The following spring, the family was still living with friends. Flood insurance was exhausted. Their house was scheduled to be razed.

✺

Before leaving Houston, I drove an hour south to see Galveston. After crossing the bridge from the mainland, I headed to the Gulf-facing side of the island, searching for something specific. I found it perched near the midpoint of a ten-mile-long sea wall that survivors had built after the 1900 storm: a small historic marker on a slant of concrete. *Original site of St. Mary's Orphan Asylum*, it read. Its metal plaque faced the street like a plea to preoccupied drivers. Its back was to the water, forever angry at the sea.

When the cyclone struck, ninety-three children and ten nuns had been inside the orphanage that stood on this spot. They faced what was probably a Category 4 hurricane, in a wooden structure at beach level, a few feet from the Gulf of Mexico. Only three of the children survived; the nuns, who tied themselves to the children with clotheslines to try to save them, all perished.

I thought about what the nuns must have felt the morning before disaster. From that grand building in a thriving city, it was probably hard to believe they were in danger. Hard to fathom that their government would ignore a life-or-death warning—or that the sea, so lovely outside their windows, would betray them. By the time they saw destruction approaching that terrible September day, it was too late.

I descended a few steps from the wall and walked the wide beach below it, stretching as far as I could see in both directions. Small waves broke peacefully on the sand. I knew that the ocean was 2 feet higher than when the orphans played here in 1900. Each year it creeps upward, faster now than before.

As I returned to Houston, I passed the Johnson Space Center. Its location on the road from Galveston suddenly felt symbolic. We've come so far. If Galveston faced the same storm today, it would benefit from modern weather forecasts, and its children would survive.

But today, Galveston Island is being swallowed by the sea. Today, every child on Earth is in peril.

☼

Only four months after the Galveston hurricane, an oil gusher erupted near Beaumont. With Galveston reeling, Houston rose—becoming the heart of the new US oil and gas industry. The city would lead in a new age, one of astounding feats in medicine, science, and technology. But also an age that would move carbon from ground to sky, and transform weather itself.

The satellites log evidence of this, year by year. They tell us that our warning systems may soon be unable to prevent tragedies on Galveston's scale. Hurricanes are becoming stronger; the next decade could see the creation of a Category 6. Their winds are intensifying more rapidly as they approach land, with higher storm surges due to sea level rise. And they move on more slowly once they strike, becoming what one journalist called "mutant sloth" storms like Harvey, which sit in place and dump rain.

Anyone can go on NASA's website and peer through the satellite's lens: at Hurricane Harvey, the Tubbs Fire, and other recent "unnatural disasters." It's a grim photo album that builds to a flurry in 2020, where an endless string of cyclones heads for the Gulf, while thousands of smoke rivers congeal over the West.

These are images of our home, under assault. They are both horrifying and mesmerizing, like an injured body in the emergency room reminding me that the laws of physics can't be escaped, no matter what our minds believe.

But when I look at these photos I also remember what the satellites can't see, at the opposite end of the visual scale: the microscopic remolding of young brains caught in such massive events. The effects of toxic stress on a mind's ability to learn and love. The trauma to children whose lives have been upended by storms and fire, drought and hunger and war. The rage and activism of adolescents who feel as if they have been handed the end of the world.

Back at my hotel that afternoon, I gazed out at the skyscrapers of downtown Houston, jutting from the flat terrain, casting a long shadow over their neighbors. I knew those companies' pumps were nearby, unearthing oil from its depths. Many oil and gas facilities had been closed and damaged by Harvey. Their toxins had swept into neighborhoods like Meyerland, and settled into the sediment that swirled around shattered homes. When I walked Lucas's street, I saw a small child shovel this tainted sediment into a plastic bucket. I knew that, as happens with wildfire smoke in my town, his pediatrician and parents would never know all the pollutants he had been exposed to, or their impacts.

But there is no mystery about oil's other toxins: NASA has made their toll clear. Every hour, more of them accumulate above us, guaranteeing the disasters to come.

When darkness fell I imagined Lucas with his family going outside to look at stars, and pointing to the satellites passing by overhead. Not yet realizing that those blinking lights are mute witnesses to a crime.

Our minds are formed by our genes and our experiences. They sometimes fail us. We ignore what we can't see or don't want to believe. We deny the consequences of our choices, even when science shows them to us as clear as day.

But we also risk our lives to help our neighbors. We get into boats and rescue families we don't know. We drive flooded streets to find a mother and her kids, muddy and alone by the side of the road, and take them in. We tie ourselves to babies and try to save them from a storm.

We build satellites, and sea walls, and children's hospitals. And we have invented the tools we need to solve humanity's most pressing problem, if we act before it is too late. We don't have to wait helplessly on the beach as catastrophe approaches; despair is exactly what the fossil fuel industry wants.

That first day we spoke, Tess sent me pictures of their house after the flood—of everything they owned in splintered, heartbreaking piles. And another photo, taken by their neighbor as they pulled away from the roof. She is sitting in the skiff with her kids and dog and the rescue team, leaving home behind.

Lucas is right beside her, looking straight ahead. At water as wide as the sea.

Chapter Four

SMALL LIFE

"During those months, everyone had it," she says in Spanish. "The neighbor who lived in front of us, the neighbor who lived next to us. It was like now with Covid, you know, day by day there was someone new who had the virus."

She is cradling her five-year-old daughter, Darah, in her arms. The little girl is sound asleep, mouth open, her small head falling back slightly. On the wall behind them, above the couch in their apartment in Hackensack, New Jersey, is a recent portrait of her. She smiles sweetly from its frame, sporting short black wavy hair and a blue polka-dot dress. Her face glows in the photo, and I recognize that look: she is a much-loved child.

"We told ourselves it was the sickness that was 'in style' at that time." Claudia looks straight at me and speaks in a steady, confident voice, never releasing her gentle hold on Darah. It began as nothing, she says. Her husband complained one morning about a rash, tiny pink blotches and bumps all over his body. Maybe he was reacting to something, they weren't sure. Three hours later he was chilled and shaking, his eyes were bloodshot, and his skin felt like fire. By the evening Claudia had a rash, too, on her arms and legs. She had been intimate with her husband the night before.

It was her first trimester of pregnancy, and Darah was the baby in her womb.

In those days Claudia worked in marketing in their home country of Honduras, where they lived with their son. She also taught marketing at the local college. Her husband is a journalist, her uncles are doctors, and her mother is a microbiologist. She knew the risk. "I always protected myself," she says. "I used the repellant, fumigated my house, and tried to be as careful as possible." She pauses, then explains: some people made her feel that what happened to Darah was her fault.

Over the next week her husband grew sicker. His muscle pains and headache worsened, and his fever wouldn't quit. She took him to the doctor, who confirmed their fear: he was fighting the Zika virus, acquired through the bite of a mosquito—an everyday pest where they lived.

The doctor told her that she and her baby were likely uninfected. It was 2015, in the early phase of the outbreak. Only a few cases of sexual transmission of Zika from men to their wives had been reported, and scientists thought such transmission was rare. No other mosquito-borne virus is transmitted sexually; the fact that Zika had ever been passed this way was startling.

Claudia's mother also tried to comfort her. Even if she did have Zika, she said, the virus might not cross the placenta of every infected woman; her odds were good. Claudia lay awake that night, hoping that she was among the spared. She followed up quickly with her obstetrician, who diagnosed her rash as allergic. Probably a reaction to that new disinfectant she'd been using, the doctor said. Everything was fine.

But in her sixth month of pregnancy, something was wrong. Claudia and her husband had gone to their regular obstetrician visit, excited about an ultrasound that would tell them the baby's sex. They were happy when the scan confirmed they were having a daughter, since they already had a son. Then, as she moved the probe over Claudia's pregnant belly, the doctor fell quiet. She finished her measurements, lifted the probe, and wiped the jelly from Claudia's taut skin. Her expression was somber as she turned to the young couple. The ventricles are fluid-filled spaces in the brain, she explained, and one of their baby's ventricles was too large. They would need to see a specialist.

Soon after, a high-risk pregnancy doctor gave them the terrible news: Darah had *microcephaly*, an underdeveloped brain. A blood test would later confirm that Claudia had indeed been infected with the Zika virus, all those months before. The specialist told them she was very sorry. Their baby was the first case of Zika microcephaly she had seen. Of course on that day none of them knew what lay ahead. Over the next two months, each time Claudia returned to that clinic, more mothers had been given the same diagnosis.

The baby's condition was obviously a shock. She and her husband cried, yet also hoped for the best and tried to prepare. Claudia made it her mission to learn what she could do to help her daughter. She searched for items that might stimulate Darah's brain or ease her care: toys and room decorations, special therapy tables and floor mats. It was in this spirit that, in her third trimester, she flew to New Jersey to visit her aunt for a week. She planned to buy some things for the baby that she couldn't get in Honduras.

Then their lives took another sudden, unexpected turn. A month before her due date, far from home in a foreign country, Claudia felt unwell. Her aunt rushed her to a Hackensack hospital. She was in preterm labor, the doctors told her, with too little amniotic fluid and a baby who already had a life-threatening condition. They had to intervene.

Darah was delivered emergently by C-section on May 16, 2016. And this little disabled girl, now snoring softly in front of me, made headlines around the world.

A Plague Among Children

Four years before the first case of COVID-19, a lesser pandemic briefly startled America. No schools locked down when the Zika virus struck the Western Hemisphere in 2015; no one wore masks or avoided restaurants. It was spread mainly through the bite of a mosquito, not breathing, and posed grave risk to only one small part of the population: fetuses developing in their mothers' wombs. Though it sickened hundreds of thousands of people in Latin America with flu-like symptoms—and sometimes a temporary paralysis known as *Guillain-Barré syndrome*—in the US Zika's reach was limited. When the virus faded in late 2016 as quickly as it had arrived, our collective memory of it faded, too. Yet some of the tiny lives it touched, like Darah's, were forever altered.

She was the first baby born in the continental US with *congenital Zika syndrome*, a constellation of birth defects that soon became notorious for its most striking feature: microcephaly. This devastating anomaly, which had spiked in a wave of Brazilian newborns the prior fall, was the world's first clue to a bizarre new health threat. Doctors soon named it and started to untangle its deeds. The Zika virus hunted young nerve cells; if it infected a pregnant mother it halted her fetus's brain growth, giving the baby an abnormally small head. Why this previously benign virus suddenly began behaving this way, seven decades after it was first discovered, is still not fully understood.

I read about Darah's birth one day in the clinic where I work, thousands of miles from New Jersey. "Did you hear?" a fellow pediatrician asked me as we stood at the nurses' station, writing in patient charts. We had been assured by public health officials that Zika was unlikely to become epidemic in the US because the mosquito most responsible for spreading it, *Aedes aegypti*, prefers more tropical climates. Yet the virus was moving north, and now plagued American territory: Puerto Rico and the US Virgin Islands had been invaded about five months earlier and were struggling to keep the outbreak contained.

Later that day I met an expectant couple for a "prenatal interview" to answer their questions and explain what would happen in the hospital after their baby was born. Had they traveled to any Zika-outbreak areas? I asked, a question that is still routine. They had, the prior winter, to a Caribbean island where Zika had been detected shortly after they returned home. The obstetrician had ordered blood tests to see if either parent had contracted the virus. "We heard about that baby in New Jersey," said the mom, who until that moment had been smiling. She nervously opened and closed our practice's brochure, which I'd handed her when we sat down. "It's scary."

It *was* scary, for more reasons than she realized. And since part of my job is reassurance, I kept those reasons to myself. I knew there was a much bigger context for the story she'd heard, not revealed in the headlines. I had been following the rumblings of scientists who study insects and disease. They were already suggesting that abnormal weather in Brazil had played a role in Zika's emergence and spread. And that we might be glimpsing a future in which mosquito-borne illnesses surge around the world.

Global warming, they warned, was literally coming back to bite us—in the form of a blood-sucking insect.

Webbed Together

To understand the connection between climate change and Darah's case, we have to zoom out from her small New Jersey apartment and see that she shares this planet with trillions of other living things. That her body is linked to the Earth not just by water and air, but by a rich sea of organisms, friend and foe, living within and around her. Many of them are being affected by rising temperatures and shifting rains; by changes in habitats and seasons.

These include various microbes that cause "infection," a word we use all the time but whose meaning reads like science fiction: invasion of the body by another life form that makes us sick. Infections are produced mostly by bacteria, parasites, fungi, and viruses, though biologists still debate whether a virus—just a bit of genetic material that needs the cells of another organism to reproduce itself—should be considered alive or not.

And it includes blood-sucking pests like mosquitos and ticks in which some of these pathogens are carried, as well as the mammals and birds in which many of them evolve. A staggering variety of life is being affected by the climate crisis, with populations rising, falling, mutating, or migrating in response to changing conditions. The effect on human illness is difficult to predict. But most experts agree: infectious disease is another growing threat of the warming world.

✺

That threat is the newest phase of a never-ending, back-and-forth war. Pathogens attack, we defend; some mutate to regain the upper hand. And while we've made incredible advances against them, COVID-19 was a reminder that they can still quickly deflate our hubris. We are the most dominant species on Earth, having conquered every continent and remodeled the planet itself. Yet like Gulliver pinned down by the tiny Lilliputians, we are often at the mercy of the smallest living and quasi-living things—who in our tale have no conscious intent at all.

It is small life that rules us. Every day in my clinic, I swim in its invisible sea. When I examine a runny nose or sore throat, or the flushed cheeks of a toddler with a high fever, I am witnessing the assault of a virus or bacterium. My distant predecessors didn't know they were battling these pathogens; not until the late nineteenth century did doctors realize that "germs" were causing most of the childhood ailments they treated. I've often wondered how they coped with watching so many of their little patients go from wellness to the grave in hours, fighting an army of baffling killers without any of the weapons that I take for granted.

While childhood mortality in the US has been greatly reduced, infections are still the bread and butter of any pediatrician's day. Most are non-life-threatening illnesses that every parent has seen: the common cold, ear infections, strep throat, croup, warts, roseola, and urinary tract infections, to name a few. But some are frightening in the speed and fury of their attack. And when those infections strike—when I have to do a spinal tap on a sick newborn who might have bacterial meningitis, or rush a child with severe pneumonia to the emergency room for fluids, antibiotics, and oxygen—I remember to be grateful for the tools I have, and to respect my tiny, menacing foes.

Now climate change is aiding these unseen enemies in two ways: by weakening us, and by strengthening some of them.

A child's immune system functions best when he is well nourished, well rested, and well loved. Yet escalating floods and droughts are increasing hunger and stress for many children of the world, lowering their immunity and making them more susceptible to infection. In the African country of Madagascar, for example, a severe drought left half the country's children malnourished in 2021, leading to higher rates of malaria, diarrhea, and respiratory infections. And after devastating floods struck Pakistan in 2022, malnutrition skyrocketed among its children, weakening their ability to fight the pathogens that surged around them—in stagnant water, in mosquitos, and in the breath of displaced crowds.

Unfortunately, natural disasters aren't the only climate-linked threat to our food supply. Dwindling bees, crop-attacking fungi, vanishing fisheries, and declining nutrients in rice and wheat are also raising the specter of widespread famine. Because infants and toddlers need 3 to 4 times as many calories per kilogram of body weight than an adult, and 2 to 3 times as much water, they are much more at risk when food and water are scarce or contaminated. They also have lower reserves of fat and depend on a steady supply of certain nutrients for growth, brain development, and to build the immune system.

That's why malnutrition not only harms them directly—already stunting the potential of millions of the world's children—but also opens a door of opportunity for microorganisms. Serious infections then increase a child's need for energy while impairing his appetite and digestion. In fact, starvation in young children is so intertwined with infection, each problem worsening the other, that it is almost impossible to separate their individual impacts.

Protecting a child from infectious diseases also requires a functioning government and stable society. Our advances against childhood ailments grew from the infrastructure of prosperous democratic countries: universities and hospitals, clinics and pharmacies, water treatment plants and mosquito-control programs, factories and roads. If environmental conditions become dire, crumbling economies and civil unrest could destabilize all these systems, undermining our ability to fight disease. It's a scenario that already plagues some poorer countries—and one that public health groups have fought for decades to reverse, not spread.

But global warming is also boosting many pathogens themselves. A 2022 study found that of 375 known human infections, only a handful have decreased due to extreme weather, while more than half have gained ground. Not surprisingly, most rising diseases fall into several large categories that are highly sensitive to climate, either directly or indirectly.

Some Like It Hot

Some *fungal infections*, for example—not a major source of human illness in the past—are surging or morphing into more dangerous forms. *Water- and food-borne infections*, mostly bacteria, are triggering outbreaks of vomiting and diarrhea as natural disasters disrupt clean water supplies, sewage treatment, and refrigeration. Other waterborne bacteria are causing skin, respiratory, and blood infections in people who wade through warming coastal areas, or floodwaters after a storm. And a brain-invading, freshwater amoeba has tragically struck a few children from Florida to Minnesota who swam in warmer-than-usual lakes and streams.

Yet two rising categories of infection may endanger humanity more than any other—and Zika is an example of both.

First, climate change is one of several factors—along with deforestation, biodiversity loss, the illegal wildlife trade, and "factory farming"—increasing the risk of *zoonotic infections*: those that evolve in animals and are sometimes still shared with them. These infections, which account for up to three-quarters of all new germs that sicken people, have always been a potent source of pandemics and epidemics, from the Black Death and Spanish flu to Ebola and HIV.

While some zoonotic diseases come from farm animals—most famously, the deadly influenza strains that emerge from pigs ("swine flu") or poultry ("avian flu")—animal-borne infections have been rising in recent decades mainly because of new pathogens from wildlife. Strangely, this has happened as wildlife populations plummeted around the world. In fact, today's children are growing up on a planet with *70 percent* fewer wild animals than were alive in 1970, part of an ecological catastrophe often referred to as the "twin crisis" to climate change, the loss of biodiversity.

These seemingly contradictory trends—more zoonotic diseases as animals disappear—have the same root cause: human disruption of animals' homes and lives. Climate change is a source of enormous disruption, driving some species to extinction while increasing our infection risks from others. One way it does that is by forcing animals to migrate. As thermometers climb, rainfall patterns change, and forests wither and burn, many species are on the move, encountering one another and people in new places, sharing the viruses they carry. Our most recent plague, COVID-19, may be an example: one group of scientists has traced its origins to the effects of climate change on bat habitat in southern China, and how the bats migrated and exchanged viruses in response.

But no category is more directly affected by Earth's warming than *vector-borne infections*—those pathogens, like Zika, that use a "vector" animal to pass between people, hitching a ride inside its gut the way we use an airplane to get from one city to another. The mosquito is by far the world's most important vector, sickening hundreds of millions of people every year. Certain ticks, fleas, biting flies, and bugs are also vectors, each of them transporting a different germ. These animals are *arthropods*, strongly affected by temperature because they are cold-blooded and tiny. And many are thriving as the climate warms: expanding their range, growing in number, biting more people, and transmitting higher loads of viruses and parasites.

As one infectious disease specialist reminded me, though: it's complicated. Each vector species has its own ideal range of temperature and humidity; they aren't all responding to global warming in the same way. And these tiny animals, like Earth's other creatures, are facing more than a shifting climate. "We are rapidly changing our landscapes all over the world," Dr. Amy Vittor said, referring to the destruction of natural ecosystems for urban and agricultural uses. A professor at the University of Florida who studies the connections between human activity and vector-borne illness, she noted that "If we're going to make generalizations about climate change and these diseases, it is exactly those mosquito species, those tick species that thrive under human-made changes—who can take advantage of a warmer, less wild, more developed world—that are going to cause us the most problems." The *Ixodes scapularis* tick that spreads Lyme disease is one of them, she said. So is the *Aedes aegypti* mosquito.

"*Aedes aegypti* loves human beings. Why? Because it thrives in urban conditions. Especially where there is poverty and water insecurity. And it not only carries major diseases like dengue and yellow fever, but has proven it can acquire new zoonotic viruses, like Zika and chikungunya, that are emerging on our warmer, less wild planet. So it's really, really opportunistic," she told me. "That's the one to watch."

The Zika crisis was a textbook illustration of what scientists like Dr. Vittor predict. The virus arose in a monkey of central Africa, was transmitted by mosquitos to nearby people, and hopped across the world in the blood of human travelers to reach countries where no one had ever encountered it before. As it stalked the Americas, both the virus and its *Aedes aegypti* host were boosted by warmer temperatures in ways that made them significantly more dangerous to people. Looking back, scientists would see that Zika struck most fiercely in hot, dry, urban places, and that those places were unusually hot and dry because of climate change.

But Zika was iconic in another way: it joined a long list of epidemics that have harmed babies and small children most. Which is why every American pediatrician was on edge when Darah was born that spring, at the start of our mosquito season. And why Zika—despite its later, sudden quiescence—is still a worry, and a warning.

The Epic Struggle

"Hello, George," I said. My face was about a foot in front of his as I waited for his wandering blue eyes to focus. His tiny hand was wrapped around the finger I'd just offered him in greeting. After a moment he found my smile, and a beat later he smiled, too—and started to coo.

"It's so fun," his mother said. She was bent over the exam table with me, stroking his hair. "It's like he's a real person all of a sudden."

She and her husband had tested negative for Zika, and their baby had been born healthy. Now he was here for his two-month checkup, doing what any normal infant his age, with a normally developing brain, should do: socializing with every person who smiled at him.

Though George had dodged a missile during the pregnancy, I still felt uneasy. A month before his visit, in July 2016, Zika had erupted in Miami, Florida; over 1,470 locally acquired and imported cases, including 300 pregnant women, would eventually be reported in that county. I worried about the young families there and in my own practice, many of whom visited Latin America regularly. Several people with travel-acquired Zika had been diagnosed in Nevada, though no one feared an outbreak here because *Aedes aegypti* had never been found in the state. We didn't know that within a year, it would reach Las Vegas.

I ran my palm over the top of George's head—the "pediatrician's handshake"—and found that his *anterior fontanel*, or "soft spot," was normal size. His little arms flailed as he stared at my face, waiting for another smile. "You're doing great, George," I told him, and then praised his mom, too. His heartbeat was steady; his lungs were clear; his belly was round and gurgling.

He had made it through the newborn period, when bacterial infections, especially, pose great danger. But for the next few years he would still be in a vulnerable group: children under five years old. While babies and toddlers manage many common viruses well—if not, none of them could go to day care or have siblings—they are clearly at higher risk from a wide range of infections because of their immature immunity and small size.

Every pediatrician knows this like we know our own name; it's why so many epidemics of the past took their biggest toll on the youngest. "It was literally the plague among children," Noah Webster wrote of one eighteenth-century diphtheria outbreak in New England. "Many families lost three and four children—many lost all."

That past reaches into my present every day—and touched George's life during his visit. As I left his exam room, I asked my medical assistant to give him his two-month vaccines, the first set that babies receive. She went to a special refrigerator where the names of prior and ongoing "children's plagues" appear on hundreds of small glass vials: polio, measles, pertussis, rotavirus, mumps, and more than a dozen others. Their neat sorting into colorful bins gives no hint of the chaos from which they arose. She selected four vials, drew them into syringes, and gave George the shots and oral drops that would prevent several life-threatening infections, including the once-feared diphtheria.

The risks to young children are why most vaccines are given in the first eighteen months of life. It's also why the health of this age group is always used as a benchmark of a country's progress. A century ago in the US, for example, 40 percent of all deaths were children under five, mostly due to infectious diseases. But by 2018—thanks to vaccines and other public health advances like clean drinking water, antibiotics, pasteurized milk, and mosquito control—only 2 percent of American deaths were babies and small children, and none of the top ten causes were infections.

Humanity's success against pediatric infectious disease is one of the great triumphs of civilization; hundreds of millions of lives have been saved. Whenever I am overwhelmed by the scope of climate change or feel enraged at those who knowingly inflicted it on today's children, I remember that from the eighteenth century through today many brilliant and good people—scientists, doctors, nurses, and public servants—have devoted themselves to children's welfare and achieved what had previously seemed impossible.

Yet huge challenges remain, as almost 2 *million* children under age five—mostly in developing countries—are still killed every year by pneumonia, diarrhea, malaria, and neonatal sepsis. We have the tools to save many of these children; poverty, indifference, government dysfunction, and strife are often as much to blame for their deaths as infection.

Now the search for better prevention programs and treatments, especially in poor countries, is even more urgent. As pediatric infectious disease specialist Dr. Peter Hotez has said, "We are in a bit of an epic struggle between developing these new technologies fast enough versus the rise of disease due to climate change" and other social and environmental disruptions.

Mosquito-borne illnesses are a battleground in that struggle for good reason: no threat in history has taken more children's lives. And vaccines against these diseases have, until recently, been few and far between.

The Most Dangerous Animal

Darah's father didn't pay much attention to the mosquito that changed his life. He swatted at it and moved on with his day. Not until later did he realize that a tiny, fragile insect had inflicted grievous harm on his family. That he and his wife and daughter had joined the roughly 700 million people sickened each year by a pathogen carried by mosquitos.

Other victims were spread around the globe; too many for him to know. In my clinic I encouraged parents to spray their kids with insect repellant, but few were concerned about disease. Most still saw the bugs' itchy bites as just a benign rite of summer and childhood.

They were almost always right; apart from occasional cases of West Nile, we rarely see mosquito-borne disease here in the high desert. But our ease is a luxury of place and time. The lowly mosquito—so easily cursed, slapped, and forgotten—is humanity's most enduring foe, responsible for more illness and death than any other cause. The pathogens it carries are thought to have killed *52 billion* people—nearly half of all those who have ever lived. Most of them were babies and toddlers.

In other words: nothing has killed more children than the mosquito. Not war or famine; not polio, diphtheria, smallpox, cholera, or plague. Though mosquito-control programs have greatly blunted the insect's impact, almost a million people continue to die annually from mosquito-borne illnesses across the globe. No other animal inflicts such a heavy burden.

Roughly half of today's mosquito-linked deaths are children under five who succumb to malaria, caused by a microscopic animal parasite. Malaria has also been responsible for most of history's toll. But mosquitos carry multiple other germs as well: a dozen viruses besides Zika, as well as the roundworm parasites that cause *filariasis*, a disabling tropical infection that obstructs the lymph system and leads to huge, fluid-filled limbs. The parasite or virus is pulled into the mosquito from one warm body, and disembarks into the next, via the mosquito's *proboscis*, the needlelike mouth it drills through our skin.

Not just any mosquito will do. Each of these pathogens has a preferred "airline" among the insect's 3,500 different species, only a few of which carry disease. Malaria travels with the *Anopheles* genus of mosquitos; the *Aedes* genus carries Zika, dengue, chikungunya, and yellow fever; West Nile flies with *Culex*. All three groups live in the US but vary slightly in their ideal habitat and climate, which determines where they are found—and where disease could follow.

Like many infections, mosquito-borne diseases declined precipitously in the twentieth century. Outside of sub-Saharan Africa, global mortality from malaria alone dropped 90 percent due to bed nets, window screens, pesticides, insect repellants, reduction of breeding sites, and antimalarial drugs. To this day, in many US counties, mosquito-control programs work to contain the insects and track their pathogens, which have been virtually eliminated as a major threat. But three zoonotic, mosquito-borne infections—West Nile, Zika, and chikungunya—are recent arrivals in the US; the last two emerged in just the last decade. Clearly our efforts to control mosquitos must be redoubled as the risk of their diseases grows again in many places.

One reason for that growing risk: milder winters, hotter summers, extremes of rainfall, and shifting winds are changing the insect's range. Some parts of the world are becoming too hot or dry for certain types of mosquitos, and every species has a limit of what it can survive. But for now, the "tropical illness zone" around the equator is enlarging, reaching higher elevations and farther north. Areas that would have previously been too cool for tropical species are suddenly open to them. In Florida alone, ten new types of invasive mosquitos have been identified since 2000.

As Dr. Vittor noted, one of the most important disease-carrying species, *Aedes aegypti*, is spreading fast; it is now found in twenty-six states. Unless we can find solutions, such as vaccines and mosquito-control programs, to stop it—and get the public to accept those solutions—the next wave of Zika may wash across large swaths of the US that were previously untouched.

Of course, Zika won't be the only disease that *Aedes* brings.

Breakbone Fever

The little girl looked so sick in the photo—glassy-eyed and limp, lying in a hospital crib with an IV in her arm—that for a moment I didn't recognize her. "I'm sorry, when did you say you got back?" I asked her father.

"Two days ago," he said, retrieving his phone from me and scrolling through it again. "No one else in the family was sick, just her. But the children's hospital was full. Here, look."

On his screen I saw a brightly lit hospital hallway crowded with gurneys. A child was lying on each, some with parents alongside; a hand-painted Elmo tried to cheer them from a mural on the wall. Her dad swiped the image aside; a young, dark-haired woman in green scrubs was suddenly smiling at me. "That's her doctor, on the day we were discharged," he said.

Then he swiped again. "And this is the room she was in." Two neat rows of cribs, each shrouded by white mosquito netting, lined the left and right sides of a long room. A tired-looking nurse stood in the aisle between them, holding an oxygen tank. Except in historical photos, I'd never seen this type of barracks-like care in a US children's hospital. And I'd never needed mosquito netting in the places I'd practiced.

A hazy face peeked through one of the nets: it was Emma, standing up in her crib and looking at the camera. I turned from the screen and glanced down at her. The top of her head was only a little higher than her dad's knee. She was wearing a yellow cotton dress over her pull-ups, gripping a plastic giraffe with both hands. She stared at her two older brothers as they broke into an ill-timed game of tag in the cramped exam room. "Hey, that's for outside, sit down," their father said, and they did.

They had gone to Honduras to visit the kids' grandmother, a trip they take every year. Emma was almost three years old; I'd known her all her life, and she'd always been healthy. But in the second week of their trip she developed a high fever and lost her appetite. When she started moaning and refused to walk or drink, her parents took her to the local hospital. There, a blood test showed that she had contracted the dengue virus.

She was far from alone. It was 2019—not long after the Zika crisis—and another mosquito-borne illness was pummeling the Americas. Dengue would infect more than 3 million people across the hemisphere by the end of the year, but Honduras was especially hard-hit. The small country suffered its worst dengue outbreak in fifty years, overwhelming its clinics and emergency rooms. By the time Emma was diagnosed, babies and young children were already making up the largest share of hospital admissions.

She was dehydrated when she arrived on the pediatric ward, and cried that her head and belly hurt. Dengue is known as "breakbone fever" because of the deep pain it causes, and Emma's father grimaced as he recalled her symptoms. The young man also felt guilty: they'd inadvertently brought their kids to the center of an epidemic. He didn't realize that the epidemic had also been brought to them.

※

Of all illnesses carried by mosquitos, dengue is responding most dramatically to climate change. The disease has become thirtyfold more common around the world in the last half century, and the number of countries afflicted by severe outbreaks has grown from nine to over one hundred. Dengue now strikes more than 400 million people a year, of whom about 40,000 succumb to bleeding, organ failure, and *dengue shock syndrome*, caused when blood vessels become "leaky" in the fight against the virus. Historically, almost 95 percent of global dengue cases have occurred in infants and children, who are 15 times more likely than adults to die from the infection.

Dengue fever is not a new disease; it has been a plague of the tropics for centuries. The virus was largely wiped from the Americas by a massive mosquito-control program in the 1940s but returned in the 1970s when those efforts waned. Most cases in the US are travel related. Over 5,000 people carried dengue into the country between 2010 and 2017; it's the most common cause of fever in Americans returning from Latin America and Asia. If these travelers go home to a community with *Aedes* mosquitos, they can start an outbreak—which is exactly what's happened in Hawaii, Florida, Texas, and New York over the past decade. Like Zika, the 2019 dengue outbreak eventually reached Miami, sickening over 400 people in Florida before it was contained.

We can expect the number of states, and cases, to grow. Scientists who have modeled the impact of climate change and urbanization on dengue found that 60 percent of the world's population may be at risk of the disease by 2080—a scenario with profound implications for children.

Dengue has four different strains, or *serotypes*. A person who recovers from infection with one serotype has much higher odds of severe disease if they catch another. Infants and young children are most likely to develop severe disease, which is usually signaled by bleeding from the nose, mouth, or gut and falling blood pressure.

Infected pregnant women are very vulnerable as well. Dengue infection raises the odds of stillbirth and greatly increases a pregnant woman's risk of mortality. If she is infected with the virus at the time of birth, her newborn can develop severe dengue within days.

Infection during birth is also a risk with chikungunya, a related *Aedes*-borne virus that reached the Americas in 2013—about the same time as Zika—and that is also exploding in the world's tropical regions. About half the newborns who catch chikungunya from their mothers develop *encephalopathy*, infection of the brain.

Because of these dangers, all pregnant women and children who travel or live in areas of endemic mosquito-borne illnesses should use an insect repellant deemed effective and safe for anyone over two months' age, such as those that contain DEET (diethyltoluamide) or picaridin. Unfortunately, such products are costly, inaccessible, or impractical as a means of daily protection for many low-income people in the world.

Which is why mosquito-control programs and vaccine research are currently our main hopes against these illnesses. Scientists have chased a dengue vaccine for decades, but the complexity of the disease and our immune system's response to it have long frustrated their efforts. A breakthrough came in mid-2021, however, when the CDC recommended a dengue vaccine for the first time. Called Dengvaxia, it can be given to children nine to sixteen years old living in communities where the virus is common, but only after they've had a prior dengue infection. No vaccine is yet available for babies and younger children like Emma, the age group most at risk of severe illness as dengue spreads.

I knew the scenes Emma's father had shown me would likely become more common in the years ahead. And not just because *Aedes*'s territory is expanding. The life cycles of both the mosquito and its pathogens are also changing in ways that boost the spread of disease. Nothing makes that clearer than the story of the Zika virus, and the journey it took to reach Darah.

The Path to Claudia's Door

Zika erupted from northeastern Brazil during one of the worst droughts in the country's history. At first glance, that made no sense. Everyone knows that mosquitos need water to breed, and swarm after a summer rain. So why did Zika flare in a place that was abnormally dry?

The mystery began in the state of Pernambuco, on the country's Atlantic coast. In October 2015, health officials there had alerted the World Health Organization that twenty-six microcephalic babies had been born at its hospitals in under three months, a much higher rate than is normal.

Not yet sure of the cause, investigators quickly descended on the hardest-hit neighborhoods, aiming to trace the outbreak to its roots. They found that many of the babies were from poor communities where running water wasn't always available. Residents coped by storing water, usually in large open containers. As the drought progressed and natural pools dried up, mosquitos were increasingly drawn to these barrels and troughs, which dotted thousands of yards and homes.

Drought and poverty had thus concentrated mosquitos in places where warm bodies were plentiful and window screens were rare. And the urban-loving mosquito best suited to this circumstance was none other than *Aedes aegypti*: the long-feared bearer of dengue and yellow fever, now able to carry a new menace, Zika.

Of course, a mosquito without a germ in its belly is nothing but a nuisance. The Zika wave would not have hurt thousands of babies in the Americas—might never have reached Darah in her mother's womb—if the virus itself had not found a route, years earlier, to sneak into Brazil.

Later analysis of the virus's genes showed that it had likely stowed away in the blood of someone attending a 2013 soccer event in Pernambuco after traveling from French Polynesia, where a milder form of Zika had recently been reported. Once it reached Brazil the germ disembarked, via mosquito or sex, into a population of millions—all of whom lacked any immunity to slow its spread.

Its opportunities were endless. And all around it were environmental changes very much to its liking. Not just that record drought, which brought mosquitos and people together. The years of the Zika pandemic were also abnormally hot, both in Latin America and worldwide; 2016 was then the warmest year in human history. This, too, profoundly affected the virus as well as its mosquito host.

At these higher temperatures, the *Aedes* mosquito matured faster, bred sooner, and survived longer. Its population quickly rose. The insect also became hyperactive, flying farther, resting less, and biting more people. *Aedes* had responded to the warmer world like a fire doused with gas.

Once inside the mosquito, those warmer temperatures drove the virus from the insect's gut to its salivary glands in less time, meaning it could be transmitted to another person sooner. And the rate of Zika's doubling sped up, increasing the amount of virus the mosquito was carrying, making it more likely to infect anyone bitten. Since there was a chance of error every time the pathogen copied itself, faster replication also increased the odds of a mutation—such as the one that turned Zika into a nerve killer, and ultimately harmed Darah.

In all these ways, heat and drought had accelerated and magnified—perhaps even triggered—the Zika crisis. They were fueled by global warming directly, but also indirectly, via an extreme El Niño, a phenomenon in which the surface of the Pacific Ocean becomes unusually warm, altering the circulation of moisture and heat around the world. Severe El Niños occur more often now because of climate change; the one that emerged in 2015–2016, right after the Zika virus had colonized Brazil, drove global temperatures to an all-time high and worsened the country's drought even further.

But one other factor, human-caused and intertwined with climate change, had also fueled the crisis.

For the Zika virus to migrate west from where it first struck, it had to cross the vast Amazon rainforest. Even in the era of airplane travel, this famously dense, sparsely populated jungle, covering 40 percent of South America and all of western Brazil, should have slowed the virus's spread. That's what happened when the dengue virus returned to Brazil in the 1980s and then took thirty years to spread across the country. But the Zika virus was able to hopscotch across the wilderness like a stone on a pond, touching down in patches where the forest had been felled. That's because anywhere the Amazon was cleared, the *Aedes* mosquito thrived.

Deforestation is a major contributor to global warming; the loss of tropical rainforests accounts for 20 percent of the planet's greenhouse gas emissions. But deforestation has another, lesser-known impact. Between 1990 and 2016 it caused a severalfold global increase in outbreaks of zoonotic and vector-borne infections. Why would turning a plot of trees into a ranch, farm, or housing development cause nearby people to get sick? It turns out that when lush, diverse forest ecosystems are replaced by urban and agricultural landscapes, the mix of animals in the area shifts toward species of rodents, bats, mosquitos, and ticks that are more likely to carry human diseases—the opportunists, like *Aedes aegypti*, that Dr. Vittor spoke of. And of course, transmission to people and livestock is easier, because they are now nearby.

Climate change, deforestation, and poverty had sped a dangerous virus to Claudia's door. By February 2016, Honduras announced that multiple pregnant women there had tested positive for Zika. That same month, WHO investigators published their analysis: Zika was responsible for the microcephaly epidemic that had erupted from Pernambuco and was now sweeping across the Americas.

When Darah was born three months later in New Jersey, I doubt that her doctors thought about deforestation or the drought in Brazil. Invisible webs connected the baby in front of them to the rainforests of Africa, Polynesia, and South America; to life as majestic as a centuries-old tree in the Amazon, and as bothersome as a mosquito. I'm sure they didn't note in her chart what now seems clear: that her brain had been injured, at least in part, by global forces we've unleashed upon ourselves.

Camron

Camron remembers only a few, disconnected scenes. He got up one morning and told his mom he didn't feel well, then vomited. She felt his forehead and took his temperature, then put him back to bed, where he stayed for the next five days. "I could *hardly even watch TV*," he tells me.

He glances up at his parents, sitting on either side of him, to check that he got that right. They're trying to describe the sickest he's ever been, the prior spring when he was seven years old. "He would stand up, walk a few feet, then lay down and sleep wherever he was, in the hallway, it didn't matter," his mother, Stefanie, tells me. "His legs were like Jell-O."

The slight, sandy-haired second grader listens carefully to the details of his own story. "He was in rough shape," his father, Kevin, adds. "He had a sore throat, bad headache, and said his eyes hurt. That was his main complaint, through the whole thing."

The illness hit in May 2020, early in the COVID-19 pandemic. His parents took him to his pediatrician, near their small town of Solon, Iowa. The doctor examined him and ran some tests, then called them with good news: he did not have COVID-19. Watch him closely and give him fluids, she said; it was probably some other virus.

Just a few hours later, Kevin was helping Camron to the bathroom when he noticed a large, circular rash emerging on his son's upper thigh.

"We never did find the tick," his mother says.

Mosquitos like *Aedes aegypti* aren't the only blood-sucking animals thriving in a warmer world. Ticks—which are not insects but *arachnids*, like spiders—are also flourishing. While mosquitos are the most important vector on the planet, in America it's the tick that rules, spreading dangerous bacterial and viral infections into ever more areas. It holds the top spot here largely because of the effects of global warming and land-use changes on a single disease: Lyme.

Named after a Connecticut town where it was discovered in the 1970s,

Lyme disease may now infect nearly a half million Americans a year, according to insurance claims. Because some doctors diagnose Lyme based on symptoms, rather than a blood test, the exact number is unknown; fewer than a tenth of those cases are both confirmed by a lab and reported to the CDC.

Camron was one of those documented cases. When more softball-size red circles appeared on his torso, he and his parents returned to the clinic and showed them to the pediatrician working that day, Dr. Hao Tran. She ordered a blood test to look for antibodies against *Borrelia burgdorferi*, the spiral-shaped bacterium that causes Lyme. It was positive.

Kevin had a pretty good idea where his son had picked up a tick. Like many of us, the family had coped with the pandemic by going outdoors as much as possible. Kevin had explored the forests and fields of Iowa since he was a kid; he knew to dress Camron in a long-sleeve shirt, tuck it in at the waist, and pull his socks over his pantlegs. They stayed on trails, avoiding the tall grasses where ticks wait for a warm body to brush by, and applied insect repellants that contained DEET. Still, the tiny tick, no bigger than the head of a pin, snuck in.

When Kevin was his son's age he would have been very unlikely to face the same diagnosis. Solon is in Johnson County, which in recent years has had the most Lyme cases in Iowa. But two decades ago the disease was extremely rare here; it has increased twentyfold since 2000, far faster than the population has grown. Across Iowa and the US, Lyme cases tripled in the same period.

Like Zika, Lyme is both a zoonotic and vector-borne disease, and, as Dr. Vittor observed, is being spread by species that flourish in human-altered landscapes. The black-legged or deer tick, *Ixodes scapularis*, feeds on a variety of animals. But it usually acquires *Borrelia* from only one of them: the white-footed mouse, which does well in the "fragmented forests" created by human development.

As it hooked on to Camron, the tick injected saliva loaded with *Borrelia*, which then used its corkscrew shape to drill into nearby skin. In most people, this early, *localized phase* of Lyme is signaled by a distinctive bull's-eye-shaped rash that appears around the bite. Called *erythema migrans*, it grows larger over several days, spreading outward like ripples in a pond.

Roughly a quarter of Lyme patients, though, never get this famous rash. By the time Camron saw Dr. Tran he was in the *disseminated phase* of illness, from one to four months after the bite. The bacteria had now entered his blood and were traveling through his body, triggering fever and pain. His multiple rash sites were caused by bacteria seeding the skin; if anyone had biopsied these areas, they would have found *Borrelia*.

Dr. Tran knew that at this phase the bacteria can cause mayhem in organs it has infected. She ordered an electrocardiogram, or EKG, to check Camron's heart rhythm and function; it was normal. She looked for loss of strength or feeling in his limbs, paralysis of his face muscles, known as Bell's palsy, or signs of encephalitis and meningitis—inflammation of the brain or the membranes around it. Thankfully, he had none of these more serious symptoms.

She then gave Camron's parents more good news: because Lyme is caused by bacteria, it can be treated with antibiotics. Dr. Tran wrote a prescription for doxycycline, and Camron's rash and pain quickly resolved. While he tired easily for several months, he eventually made a full recovery, as do most children whose Lyme is caught and treated early.

Camron was lucky. Some patients never develop a rash and aren't diagnosed until they've progressed to the *late phase* of Lyme disease, which can be debilitating and may persist despite antibiotic treatment. Some late-phase patients suffer from mild encephalopathy that leaves them struggling with mood, memory, sleep, and thinking.

The most common late manifestation, however, is arthritis, occurring in 60 percent of these patients. A swollen, hot knee is sometimes the only symptom of the disease. In fact, this was how many of the initial cases presented. The first medical journal article about Lyme, in 1977, described it as "an epidemic form of arthritis . . . occurring in eastern Connecticut." The authors noted that this new illness was much more frequent in children, something parents in the area already knew. The outbreak had come to light when one patient's mother called the health department. Did the state realize that twelve children from Old Lyme, a small rural town, had been diagnosed with juvenile arthritis? She thought someone should investigate.

✺

Epidemiologists would eventually find thirty-nine pediatric arthritis cases in the adjoining towns of Lyme, Old Lyme, and East Haddam; they outnumbered adult cases by more than 3 to 1. Tellingly, most of the children were school age, lived along certain rural roads, and became sick between June and early fall—when tick populations were highest and the kids were on summer break.

Across the country today, children five to nine years old, like Camron, continue to get Lyme at higher rates than any other age group; they account for a quarter of all cases in the US. The disease is an example of how school-age children are more susceptible to some infections not because of their immune systems—which are strong compared to babies' and toddlers'—but because of their behavior. Polio was similar in this sense; an intestinal virus that could spread via contaminated water, it spiked when children swam in lakes and streams during warm weather.

Global warming and land-use changes are widening these risks to more children. Both the ticks and mice are expanding into parts of northern New England, the Midwest, and Canada that were previously too cold for them. Peak "tick season" starts earlier in the year and lasts longer. The tick's immature form, or *nymph*, which transmits most *Borrelia* infections, also seeks hosts more aggressively as temperatures rise. All these factors are fueling Lyme disease, and it may get worse: if the world warms 2 degrees Celsius, pediatric cases are expected to rise more than 30 percent.

Lyme is not the only tick-borne illness we have to worry about. The *Ixodes* tick also carries the rare but dangerous Powassan virus, which kills up to 10 percent of its victims and permanently disables half of those who survive severe disease. It and other tick-borne diseases—such as *ehrlichiosis*, *tularemia*, and *Rocky Mountain spotted fever*—have more than doubled, as a group, since 2004.

We don't form long-term immunity to Lyme; Camron could catch it again. A vaccine might soon be available, as several are being investigated. If any are found to be safe and effective, are approved for school-age children, and are accepted by the public, they might prevent cases like Camron's.

Fortunately, another new vaccine protects children from a different vector-borne illness—the one that has killed more than any other.

The Incalculable Toll

Caused by the infamous, single-cell *Plasmodium* parasite, malaria infects roughly 247 million people a year. While anyone can get the disease, children under five years old account for three-quarters of malaria deaths in the world. More than 475,000 babies and toddlers, almost all in Africa, were killed by malaria in 2019. That's about one child dying every minute.

Few American parents worry about their own kids catching malaria; for most of us it is a tragic but faraway scourge. Yet roughly 1,500 people are hospitalized with the disease annually in the US, and some die. As with dengue, *Plasmodium* infects these people when they visit the tropics. And as with dengue, the mosquitos that transmit it are also here, meaning that returning travelers may inadvertently start a local epidemic.

That appears to be what happened in the summer of 2023, when small outbreaks of locally acquired malaria struck Florida, Texas, and Maryland, sparking national headlines. Though only nine cases were recorded in all, they sent a shudder through the country's public health officials. It was the first time in decades that malaria had been transmitted in the US.

Anyone familiar with our history understood the significance of the disease's reappearance. While it's hard to imagine now, at one time malaria sickened American children at the same horrific rates still seen in endemic countries. In the eighteenth century a visitor to the Carolinas quipped that mosquitos and malaria made the area "in the spring a paradise, in the summer a hell, and in the autumn a hospital." The disease remained a seasonal plague across the South until after World War II, when it was finally eradicated by mosquito-control programs, better housing with window screens, and antimalarial drugs. Although *Anopheles* mosquitos are still found over most of the country, they usually fly without their deadly passenger.

Could malaria return to our shores in a bigger way? Will it get worse in

Africa, where it already breaks so many hearts? Infectious disease experts have long worried about these questions.

The answer, again, is complicated. In some parts of Africa, *Anopheles* species are declining due to extreme heat and drought. In other regions, they are expanding into elevations and regions that were previously unsuitable for them. The overall effect, so far, has been to widen malaria's range.

Analysts predict that childhood mortality from malaria may rise in some endemic areas by up to 20 percent due to climate change. The World Health Organization still aims to eradicate the disease, but can't meet that goal unless the ill effects of global warming are offset by increased public health interventions. These include giving insecticide-treated bed nets and seasonal antimalarial drugs to infants and pregnant women, who, if they catch malaria, are more likely to develop severe disease, deliver prematurely, and have low-birth-weight babies. Even as climate works against us, experience has shown, we have tools we can use to save lives.

Which makes the tragedy of malaria all the more excruciating. The disease has been preventable and treatable since the US drove it out, yet in those decades it has killed *millions* of the world's poorest children. These numbers, adding up year after year, somehow numb us to their meaning. But I once lived in a country with endemic malaria, when I served with the Peace Corps in Ecuador. I know that statistics cloak the suffering of each child they count.

The Amazon village where my then husband and I lived was a collection of two dozen wood shacks in the jungle, elevated on stilts to keep out snakes. The homes had no electricity or running water, nor windows or doors—just openings cut into plank walls, which usually stopped short of the roof. Mosquitos were a constant, and though we slept under netting and used repellant when we could, the insects still made us and the Kichwa family we lived with, including several small children, their daily meal.

Learning about malaria was part of our Peace Corps training; it was in a classroom in Quito that I first memorized *Plasmodium*'s complex life cycle. By the time we headed to our post in El Oriente, the lush eastern part of the country, I could recite all the phases of the organism as it traveled through a person's body.

Simply put: if a mosquito were to pass the parasite to a toddler in my village, it would go first to her liver cells. There it would mature and reproduce itself until the infected cells burst open like overinflated balloons. The freed parasite would then invade her red blood cells, where it would again multiply until it ruptured those cells, too. This would release more parasite to infect more red blood cells, and so on.

The explosion of *Plasmodia* from blood cells is what causes malaria's symptoms, including the disease's famous cyclical attacks. Along with the seeds of new infection, the destroyed cells unleash toxins that inflame the body, leading to fever and rigors, vomiting and pain.

I once saw a toddler with these signs in a hut deep in the jungle, where we had gone to help with a fisheries project. I watched her mother move the girl's black hair aside and put a wet rag on her forehead as she lay on a mat on the floor, encouraging her to drink water brought from the river. *Chukchu*, her father told us—the Kichwa word for malaria. I handed him a packet of rehydration crystals and explained how to use them. As we left, I looked back at her home under the Amazon's towering, buzzing canopy, hoping I might someday learn what happened to her. I knew that if the girl did have malaria and recovered, she would get short-term immunity and have less chance of severe disease the next time—the opposite of dengue, which would get worse with subsequent infection.

But the loss of red blood cells from recurrent malaria might lead to life-threatening anemia, especially in someone so young. If she developed severe disease, infected cells might stick to her artery walls, blocking blood flow; in the brain this becomes *cerebral malaria*, and often leads to death. If she survived, she might have trouble speaking, walking, hearing, or seeing for the rest of her life. This is how, child by child, malaria saps the hope and energy of the people it stalks—especially in sub-Saharan Africa, where the average child is reinfected 6 times a year.

If she did get gravely ill, her parents would carry her by foot and by bus, over machete-cleared jungle trails and dirt roads, to a medical clinic in a town a few hours away. If the doctor there suspected malaria, he could examine her blood under a microscope. If *Plasmodia* were present, he would see them easily: dark shapes lurking within her red blood cells. In the Americas either *P. vivax* or *P. falciparum*—two of the five species of *Plasmodia* that cause disease—would be most likely.

Like every twentysomething American in the Peace Corps, I knew the world divided along lines of wealth. But my time in Ecuador gave me details I wouldn't have otherwise understood. In a house that can't be closed, parents have little chance of protecting their children from mosquitos and their diseases. In a village that sometimes lacked enough food, especially protein, children are often weaker when illness strikes. When that village is remote, the children may be beyond care by the time they reach a doctor. And even if they aren't, there's no guarantee that the medication, equipment, and supplies the doctor needs will be available.

Ironically, the indigenous people we lived with had once been able to treat malaria on their own. For centuries, they knew that the disease's symptoms could be cured with an extract made from the bark of quina—the national tree of Ecuador, now known to scientists as *Cinchona pubescens*. After the Kichwa's ancestors shared their knowledge of "quinine" with a Spanish priest around 1630, it spread around the world as a treatment for malaria. In 1934 a synthetic version, chloroquine, was developed, and became crucial to US eradication of the disease. The people we'd been sent to help, in other words, had already, indirectly, helped us.

The Kichwa had uncovered a key fact: unlike viruses such as dengue, the single-cell protozoa that causes malaria can be killed with medication. It can even be prevented from causing infection in the first place, with malaria prophylaxis drugs that any traveler to the tropics, even infants, should take. Drugs that I took when I served in Ecuador.

But a problem has arisen: after decades of chloroquine's use, *P. falciparum*—the parasite's deadliest strain—has developed resistance to it in many parts of the world, including Ecuador. In some countries, this species has even become resistant to newer antimalarial drugs, like

Fansidar and Malarone. *P. vivax* has also developed chloroquine resistance, though in fewer places.

It's a story repeated over and over: when we try to kill a microorganism with chemicals, there will be members of that species who are just slightly genetically different, and are not killed. Over time, as their susceptible brethren die off, these resistant organisms become more numerous and then dominate. Doctors soon find themselves without the weapons we had relied on.

Plasmodium's growing resistance to antimalarial drugs is being compounded by a similar problem with mosquitos, who are increasingly resistant to insecticides. To be clear: drugs and pesticides have health and environmental impacts that make them less than ideal solutions. Yet they have been important weapons in this fight. Their decreased effectiveness threatens efforts to control malaria just as climate change boosts its spread.

The urgency of this situation, along with malaria's ongoing horror and terrible toll, have made finding a vaccine against *P. falciparum* the holy grail of infectious disease medicine for the past half century. The parasite's complex life cycle and ability to mutate have greatly complicated this quest. In fact, targeting a vaccine against *any* parasite—rather than a virus or bacteria—is so challenging that no one had ever done it. But a breakthrough finally came in 2021 when the World Health Organization recommended a vaccine called Mosquirix for use in malaria-burdened areas. Given in four doses between five months and three years of age, the shot reduces severe malaria by 50 percent. A vaccine that might be even more effective is currently being tested in several African countries.

WHO's decision came at a critical moment. Huge, inspiring gains have been made against malaria since 2000, when Africa's leaders pledged together to tackle the disease through every public health tool they had—mainly bed nets, pesticides, and medications—and secured financial aid from the US and other developed countries. Child deaths on the continent fell 44 percent because of these efforts, and more than 7 million lives were saved. But progress has flattened in the last several years due to climate change, the COVID-19 pandemic, medication and insecticide resistance, and other factors. Like a bomber arriving to support weary soldiers on the battlefield, a vaccine brings hope and energy back to the fight.

I am still in touch with the family we lived with in Ecuador. Since our time in the village, the mother, whose name is Maria, and a group of fellow Kichwa women have founded a midwifery service for their community. Maria told me that they watch their mothers and babies closely for signs of malaria, dengue, Zika, or chikungunya. They know that the jungle is getting hotter because of climate change, risking more premature births and lowering birth weights. And that weak babies are more vulnerable to infection.

The last time we spoke, she reminded me that fossil fuels had hurt her community in another way. Centuries after the Spaniards learned about quinine, oil was discovered beneath the Ecuadorian rainforest. A hundred miles north of her village, a subsidiary of Chevron—whose plaque hangs on the wall of Texas Children's Hospital—destroyed almost 2.5 million acres of rainforest and left behind hundreds of pits of toxic waste. The waste drains into rivers used by indigenous people for drinking and bathing; for decades, parents in the area have argued in court that their children are being sickened and killed by the pollution. Now, in a final insult, the product that was pulled from Amazon ground is warming the world, increasing the risks of heat and mosquito-borne illnesses for their babies.

News of the malaria vaccine had left me overjoyed, thinking of the millions of children who suddenly faced a brighter future. I felt inspired by those who have fought so hard against the disease, showing the goodness that human beings are capable of. But it is impossible to look at malaria's toll—or the growing toll of climate change in developing countries—and not see the low value we collectively place on a poor child's life. What kind of world might we have, I wondered, if a little girl in the jungle mattered as much as a barrel of oil.

Swallowed

Darah spent several weeks in the NICU after her birth, her name unknown to the reporters who clamored to learn more about her. "She was lovely," Claudia later recalled to one of them. "She looked to me like a normal baby. Her head was smaller, but that was all—nothing that would frighten me."

Born a month premature and with a brain whose growth had been interrupted, Darah struggled at first with the basic task of drinking milk. As was true for most tiny babies in the unit, sucking and swallowing took coordination and strength that she lacked. She otherwise did remarkably well, surprising her doctors, none of whom had been sure she would survive.

Upon her discharge, when Darah had learned to feed well enough to go home, she and her family returned to Honduras. Within the year, though, they were back in the US. Darah had begun having seizures that her doctor in Honduras could not control. To maximize whatever potential she had, her parents decided to uproot from the only home they knew and move to northern New Jersey, near Claudia's aunt. From there, Darah would be able to see specialists in both New York and Philadelphia.

When we spoke in early 2021, their immigration to the US seemed to them like the right decision. Darah was receiving speech, physical, and occupational therapy in a room of their apartment they had dedicated to her care. She regularly sees her pediatrician and a team of medical specialists, including the neurologist who successfully treated her seizures.

There was another reason their move, in hindsight, seemed wise. In November 2020 Honduras had sustained what was once an unimaginable catastrophe. Two Category 4 hurricanes, Eta and Iota, hit Central America only two weeks apart. Both stalled and dumped rain on the poor country—much as Harvey had done in Houston, but doubled. Back-to-back massive storms had struck the same struggling place; it was as if the ocean itself had fallen on them. Saturated ground sloughed off hillsides, bringing down houses, roads, and electrical lines in massive rivers of mud.

A physician working for the relief agency Project HOPE told NPR that in his many years of working in disaster areas, he had never seen anything like it.

※

Beneath the violent destruction and loss of life, a quieter crisis brewed. Water treatment systems collapsed in some cities. In rural areas, floodwaters swept through outhouses and sewage ditches and spread their contents wide. As parents and children walked through the flood to reach safety, or came back later to salvage their things, that tainted water touched their skin and splashed into their mouths. Food rotted in warm refrigerators where power had failed. Kitchens, restaurants, and shops lay buried under tons of muck or destroyed by flood, while delivery trucks were trapped by buckled and submerged roads. Many parents couldn't find clean water for baby formula or food for their children. They felt helpless as their infants and toddlers, especially, grew weak with hunger, leaving them less able to fight infection.

Meanwhile, the food and water parents *could* find was often contaminated, which led to their children falling ill with acute gastroenteritis. A doctor in one hard-hit mountain town reported at least a sevenfold increase in pediatric diarrhea cases two weeks after the hurricanes. If they became dehydrated by their illness, most children couldn't get IV fluids because so many clinics were damaged or unable to function. Those who were already malnourished became more so because of their gut infection, starting a downward spiral. And all of this happened against the backdrop of the COVID-19 pandemic, which became impossible to contain as refugees packed into rescue boats and shelters. In the chaotic wake of natural disaster, the dangers of infectious disease had skyrocketed.

Families fled north, searching for food, fresh water, and shelter. And after all the rain the area got, in yet another warmer year, parents waved away the buzzing clouds that swarmed around their children, as mosquito populations soared.

Zikas of Tomorrow

The Zika virus infected thousands of mothers from twenty-two US states and territories between 2016 and 2018, causing obvious brain and eye defects in about 5 percent of the babies who survived until birth. Over the next several years the CDC would tally more than 3,300 American infants with some type of Zika-associated abnormality. Infections in the first trimester of pregnancy, like Claudia's, were most strongly linked to microcephaly.

Research from South America showed that a quarter of infected mothers passed the infection to their babies, and a third of those cases resulted in either miscarriage, stillbirth, or severe birth defects. But Zika would also prove to cause subtler neurologic damage not evident at birth. By age two, one-tenth of US children exposed prenatally were found to have vision and hearing problems, seizures, difficulty swallowing or moving, below-average learning, or language delay. A new worry, that babies could be harmed by Zika *after* birth, grew when researchers found that infant monkeys who'd been exposed to the virus in the first weeks of life had higher rates of behavior problems and evidence of brain damage.

The Zika pandemic peaked in 2016. Then the virus suddenly died down, probably because of herd immunity. Across the Americas, its prevalence declined up to 70 percent; by 2019, there were no US cases.

But what happened in those years was likely a warning shot. In 2021, Zika flared in India. In 2022, Latin America recorded tens of thousands of cases. While dengue is still far more common, Zika will probably surge again, somewhere in the world. Without rapid cuts in greenhouse gas emissions, more than 1.3 billion previously unexposed people are predicted to be at risk of the disease by 2050.

When it returns in force, a vaccine will likely become quickly available; scientists have been studying several options. They are also searching for new ways besides vaccines to prevent mosquito-borne illnesses. In one such attempt in 2021, genetically modified *A. aegypti* mosquitos were released into the Florida Keys. All male, they pass a lethal gene to females they mate with in the wild. Since only females bite, this reduces the insect's population and ability to spread disease.

Other tools are also being tried or investigated. Researchers and mosquito control agencies are infecting the insects with bacteria that kill them, prevent their eggs from hatching, or interfere with virus transmission. NASA satellites are tracking potential mosquito habitats, and the CDC is providing software and apps to help public health agencies in their surveillance of mosquito-borne viruses.

These are our latest weapons in the endless war against infectious organisms. As climate change, deforestation, and other human activities help so many of our microscopic enemies, we'll need all the firepower we can get.

Because the most dangerous animal is not the irritating six-legged insect that we slap every summer. It's the one tearing apart the web of life in which a child and a mosquito live.

George recently came in for his five-year checkup. He was full of news about his summer. He drew me a picture of his dad, named some shapes and colors, and recited the alphabet. He was happy to be starting kindergarten in the fall.

Darah is the same age. She cannot speak or walk. Like many Zika-affected children, her eyes were damaged by the virus, so she sees only light and shadow. Some of her joints are stiff, and her hands are held in fists.

The difference between these two children is a story of luck and geography. Of struggling places like Honduras, being hit hardest and first by problems it had almost no role in causing. Of the challenges of infectious diseases in tropical climates—a climate that is now heading north.

Claudia gently opens Darah's hands and massages them every day. She is devoted, body and soul, to her little girl, spending all her time with therapists and doctors, researching experimental treatments, trying to stimulate her brain.

Some might think such efforts are pointless, and give in to despair. But Claudia's love for her child is unyielding. "She has made us better people . . . we love her unconditionally, and we treasure her."

Darah smiles wide and reacts with her whole body as Claudia hugs her. Her doctor said she would never feel emotion. That her world would never get brighter. That there was no hope of change.

Her mother refuses to believe it.

Chapter Five

THE POSSIBLE WORLD

I saw Anna again in August 2021.

I was double-booked that day, and the clinic was bustling with chaos: crying babies, talking kids, scolding moms, ringing phones. Every few minutes my medical assistant's voice would cut through the noise, asking questions and giving instructions as she brought families back from the waiting area. We had made it to early afternoon when I exited a room and bumped into a ten-year-old getting his vision checked in the hallway; he was covering one eye and squinting at the wall. I weaved past him and around a stroller, dropped off one chart, lifted another from its wire basket with a *scrrrape*, adjusted my N95, and knocked on the next door.

I'd been moving from patient to patient faster than I like. But when I entered the room and saw Anna sitting on the exam table—with her worried father beside her and dark sky in the window behind them—I slowed down.

Here we were again, in a city choking on smoke and a clinic filled with coughing children. It had been eight years that week since Anna had come in with her mother during the Rim Fire. Eight years since, as a baby, she had first shown me that climate change was hurting my patients. The hottest years in human history were all she knew.

I'd seen her many times over those years for checkups and sick visits; her asthma often flares during smoke season, and her parents always worry. "What's going on, Anna?" I asked as I closed the door. She looked back at me with those big eyes, peeking above her mask.

Her body was thin, and her olive skin seemed pale. She wore well-scuffed light-up tennis shoes and had a scraped elbow, signs of an active playground life. But in the clinic she barely moved, just watched carefully. Still my small judge, unsure of her verdict.

Anna glanced at her father, who said, "You can tell her," and nodded encouragingly in my direction. She lifted one arm and pointed at the window. I saw that she was breathing fast.

And then, before complaining about anything or telling me why she was there, she said: "It will get better."

Her young face stared at me expectantly. Was she talking about her lungs, the fires, the world? Overwhelmed by weeks of smoke and the Delta wave of Covid, I went blank. Then I followed her finger and looked outside, at a city shrouded by darkness in the middle of the day.

Only 50 miles northwest of that exam room, the Dixie Fire was devouring a huge swath of forests and towns. The Caldor Fire was raging less than 80 miles southwest, threatening the city of South Lake Tahoe. Thick smoke from both was converging, like two toxic rivers, in Reno. It would replace our air from mid-July to mid-September, and then come in waves for several more weeks into October, depending on the wind.

By the day of Anna's visit, some of my patients' homes or schools had already burned down. Other families I care for would be evacuated multiple times. Some would worry about a dad on the fires' front lines, or a mom flying a tanker of fire retardant. Almost every day a parent told me that this year was the final straw; they were moving away to protect their kids from the smoke. And all around us, familiar, once-beautiful landscapes were being transformed into a charcoal expanse, dotted with the crumpling black skeletons of buildings and trees.

"You're right," I promised her as I turned back. "The smoke always clears."

Where We Stand

Taking care of children is to be constantly reminded of time. As I listened to eight-year-old Anna's lungs, I was struck by how much she had grown since the Rim Fire, and all that had happened since. Her world is hotter and more dangerous now because of global warming. Every child whose story appears here has experienced some version of the consequences, depending on the place.

But her growth wasn't just a marker of years gone by. It was a signal of the years to come. She has decades of life in front of her, and part of her parents' and my job is to protect her potential future. The question is not how much worse climate change has gotten in the past decade, or how many children have been hurt. It's what I and everyone else will do about it in the next.

Anna will outgrow pediatric care at the age of eighteen. She'll reach that age milestone in 2030, an important date on climate scientists' calendars. By that year humanity must cut its carbon dioxide (CO_2) emissions to 45 percent below 2010 levels (or about 50 percent below 2020 levels) if we are to keep the eventual peak of global warming near 1.5 degrees Celsius. In other words, this decade—the remainder of her childhood—will decide Anna's future. It will determine the boundaries of the possible world.

That "45 percent by 2030" goal comes from a 2018 UN Intergovernmental Panel on Climate Change (IPCC) report which, when released, led to headlines that we had "only twelve years left to save the world." Its authors hoped that the urgency of the threat and the clarity of its solution would trigger an immediate global response.

Unfortunately, the world moved in the opposite direction. A 45 percent cut from 2010 levels would drop CO_2 emissions to 17.9 billion metric tons (or gigatons) per year by 2030. But in 2023 those emissions climbed to an all-time high of 40.9 gigatons, most of which will linger above us for centuries. Methane, the second-most important greenhouse gas, also rose.

This wrong-way trend means that we're using up Earth's "carbon budget" faster than hoped. In 2020 the IPCC estimated that to have a

50–50 chance of only 1.5 degrees of warming "with no or limited overshoot," we could not add more than 500 gigatons of CO_2 to what was already in the atmosphere. By 2023 we'd used at least a quarter of that. If global emissions peak by 2025 and then steeply drop, however, we still have a shot at staying within the 1.5-degree carbon budget.

The US goal is slightly different. Although America has contributed more CO_2 to the atmosphere than any other country, its emissions peaked in 2005 at 6 gigatons per year and have fallen since, largely due to the retirement of coal plants for natural gas and wider adoption of wind and solar power. Here the aim is to cut emissions to 50 percent of 2005 levels by 2030. The programs funded by the federal Inflation Reduction Act (IRA), which passed in August 2022, were designed with this goal in mind; they are expected to reduce US CO_2 emissions roughly 40 percent from 2005 levels over the next decade.

That's still not enough on our part, and other countries must also meet aggressive timelines, especially China, whose annual emissions now significantly exceed those of the US. But there are plenty of encouraging trends. The price of solar and wind energy, as well as batteries, has plummeted, making nonpolluting systems cheaper than fossil fuels in most cases. China is expanding green energy at an astonishing pace and meeting its goals ahead of schedule. And although the world's carbon dioxide emissions reached record levels in 2023, so did renewable power, which generated more global electricity than ever before; in the US it currently provides 22 percent of our electricity.

So: worldwide, greenhouse gases are being pumped out at the highest rates ever. The concentrations of those pollutants in the atmosphere are also at their all-time peaks, with CO_2 at around 420 parts per million—higher than at any point in the past 4 million years. Scientists estimate that the world now has only a 5 percent chance of keeping warming below 2 degrees Celsius. It is impossible to overstate the emergency we are in, and how muted the world's response is relative to the threat.

And yet the promise of better and cheaper renewable technologies, coupled with worldwide activism and government initiatives in the US and elsewhere, is giving us the most realistic hope of a global energy transformation that we have ever had. Millions of children's lives, including Anna's, depend on how quickly that transformation unfolds.

One Child in the Decades Ahead

Cutting emissions in half this decade may seem impossible. But in 2022 a group of Stanford University researchers detailed how most of the world's countries could, in fact, exceed that goal, and replace *80 percent* of their current fossil fuel–burning energy sources with proven, low-cost renewables by 2030. They calculated that a full transition to green sources would save these countries trillions of dollars and create millions of jobs. Their analysis made clear: our problem is not a lack of solutions, but a failure of political will.

What would Anna's future look like if we found that political will, and got on track to meet the "45 percent by 2030" goal? The IPCC recently published a temperature curve showing what we could expect in that possible world.

It shows that, even in this best-case scenario, Earth's temperature will climb by about a tenth of a degree Centigrade per decade, and Anna will inhabit a warmer planet than humans have in the past. By the time she graduates from high school, global warming will have risen from its current 1.2 degrees to slightly more than 1.3 degrees over preindustrial levels. By her twenty-eighth birthday—when she may be married and have children of her own—warming will have reached about 1.4 degrees.

She will turn thirty-eight in 2050—the year by which scientists tell us we must reach "net zero" emissions. At that point, if we have succeeded, human activities will no longer be adding to the level of carbon dioxide in the atmosphere; the amount we pollute will equal what oceans, forests, and soil can absorb.

On this optimal curve, Earth's warming will hit 1.5 degrees in 2060, when Anna is forty-eight. And when she is getting ready to retire and enjoy her grandchildren in 2075, global temperatures will finally, slowly, start to fall.

To clarify: this scenario is based on the *trend line* of global temperatures. As we saw in 2023, when an El Niño event on top of global warming spiked temperatures to record heights, some years may come close to or breach 1.5 degrees of warming. But there will also likely be years ahead that are below that level. What matters most is the overall trend. If we rapidly reduce emissions, that line—similar to a moving average over time—should not hit the 1.5-degree threshold for decades, according to the IPCC. Of course, if we don't cut fossil fuel use quickly, the trend line will cross 1.5 degrees much sooner—perhaps as early as the 2030s.

There is an important caveat, too. The latest climate science shows that if we stopped adding carbon to the atmosphere, Earth would stop warming in *only three to five years.* Though the planet wouldn't cool immediately, a pause in warming would give civilization much-needed time to adjust to rising seas and weather extremes. If we reached this goal sooner than is currently predicted, it could lower the Earth's ultimate peak temperature, saving more lives, ecosystems, and human-built structures. The faster we eliminate fossil fuels, in other words, the better our kids' future will be.

When Anna is an adult in the best-case scenario, the world around her will look, smell, and sound very different than ours. When she walks out her front door and down the street, the cars and trucks driving by won't leave a cloud of exhaust or rattle her with the rumble of engines. They'll all be electric and self-driving, zipping silently and cleanly past. When she gazes downtown, the planes she'll see taking off and landing at the airport will also run quietly on batteries or use new, carbon-neutral biofuels.

As they do now, electric vehicles will plug in at home, work, or roadside charging stations. But unlike now, 100 percent of the electricity for them and everything else will come from solar, wind, hydropower, geothermal, and other clean sources. Large power plants run by utility companies will store that energy in giant batteries to maintain a steady supply regardless of the weather. In some neighborhoods, batteries and solar panels will be deployed in microgrids that serve a group of homes; some homes will be off the grid altogether, using their own rooftop solar panels with a wall-mounted battery in the garage.

When Anna gets to work, or comes home at night, the buildings she enters will be warmed and cooled by electric heat pumps instead of furnaces and air conditioners. If air conditioners are still in use, they won't use today's dangerous coolants, which emit powerful, eternal greenhouse gases. When Anna cooks for her family, it will be on an electric stovetop using magnetic induction technology; when she runs bathwater for her children, it will be heated by electricity.

Put simply: all electricity will come from nonpolluting sources, and everything we now power with oil and gas will instead plug in.

The city will also have taken steps to cool itself in a hotter world. A canopy of trees, in parks and yards and along streets, will calm tempers and temperatures and suck up carbon dioxide released in the past. Vines and mosses will cover some buildings, and rooftop gardens will grow on others. Car-free zones downtown will provide space for people to gather, walk, and talk. Playgrounds will be made of wood and shaded by fabric sails. Streets will be painted light gray; some may wirelessly charge the electric vehicles that drive over them. The materials used in new construction will absorb less heat. Solar panels will cover and cool parking lots, bike paths, and irrigation canals, where they'll also prevent evaporation.

Anna's diet, too, may have changed. It will likely be richer in plants, reducing the risk of obesity, diabetes, and cardiovascular disease. Her family's food will come from farmers who treat soil and rotate crops in ways that absorb more carbon dioxide, and ranchers who raise livestock in more sustainable ways. Lower-meat diets will help alleviate world hunger and lessen the risk of zoonotic diseases from "factory farms" that cram animals together in small spaces. They will also reduce the greenhouse gas emissions and deforestation associated with animal agriculture, especially cattle ranching. Palm oil, now nearly ubiquitous in processed foods, will have been replaced by products that don't destroy rainforests. And after dinner, her kids will scrape food waste into a backyard composter so it doesn't produce methane in the local landfill.

The great forests of the West will be scarred by drought and fire. But megafires will be rarer because solar-powered microgrids and buried electric lines in mountain towns will have eliminated the aboveground wires that spark so many of today's blazes. Video and satellite surveillance systems will spot lightning fires early and send crews to contain them. And California will have spent billions on better forest management, recognizing that while climate change is the primary driver of the fires—because it kills trees and dries them into tinder—the accumulation of dead underbrush from decades of fire suppression has made things worse.

In this possible world, urban smog will be a thing of the past. City air will be as clean as the air in the countryside. If Anna lives in a heavy-traffic neighborhood, her children won't face more risk of asthma or stunted lungs, because cars won't pollute the air. Coal- and gas-fired plants, with their toxin-churning smokestacks and methane-leaking pipes, will be gone. Someday, she may show her children photos of the world she knew: of brown air blanketing our cities, and oil-soaked beaches with dead fish and birds; a world where millions died every year from air pollution. They will not understand why anyone fought for that world.

Because it is the current world, on its current path, that demands that children like Anna—and Liam and Ruby, Cody and Lucas, Shani and Darah—sacrifice. It demands their trauma and illness and disability and even death to save an industry that had its time, and now must go.

Earthquake

It may seem strange that a pediatrician is talking about electricity, cars, farms, and garbage dumps. Doctors don't typically give prescriptions for infrastructure and agriculture. But green energy and sustainable living are as important to my patients' welfare as car seats, medications, laboratory tests, and vaccines. Again, children's bodies and minds are inseparable from the atmosphere. Its health and theirs are linked. The toxins we dump upward, and the steps we take to end that practice, will hugely impact their lives.

In the possible world I imagine, caring for Earth's air as if it were part of ourselves will be second nature. If that seems far-fetched, consider a crisis of the past: the cholera pandemics of the nineteenth century, which killed 5 to 10 percent of the population in some American cities. The disease spread because at that time—though it seems incredible to us now—people did not understand that they should not let human waste drain into the same river they used for drinking water; that this was how an unseen organism, the *Vibrio cholerae* bacteria, was sickening so many.

Historians may one day marvel that it took us so long to understand a similar lesson when it comes to air. Today's green energy technologies are the equivalent of the drinking water and sewage treatment plants that now prevent outbreaks of cholera and other waterborne diseases in developed countries. Public water systems are among the greatest health advances in history, but the benefits of renewable energy will be even greater. In fact, the potential health impacts of the "green revolution" led one prestigious medical journal to call climate change "the biggest global health threat" *and* "the greatest global health opportunity" of the twenty-first century. Because even if fossil fuels were not threatening us with climate catastrophe, transitioning off them would yield incalculable benefits for our lungs, hearts, and minds—preventing as many as 4.5 million pollution-related American deaths by 2100.

We already have the technology and knowledge needed to make that transition. The possible world of Anna's future does not require any astonishing new breakthroughs. Like tectonic plates moving beneath our feet, the country has been gradually shifting toward it for years. And now an earthquake is coming.

That's largely because of the recent passage of the Inflation Reduction Act. Built on lessons from past climate legislation failures, the IRA relies not on "sticks"—mandates, regulations, and penalties—but on "carrots": tax incentives that encourage the adoption of green technology. Billions of dollars are devoted to making our country the industrial center of the clean energy revolution by boosting the manufacture and installation of wind turbines, solar panels, and battery storage. The act funds home decarbonization programs so homeowners can switch out gas-burning appliances for electric alternatives. It subsidizes the purchase of electric vehicles and expands charging stations. And in the spirit of New Deal programs that helped farmers conquer the Dust Bowl nearly a century ago, the IRA supports voluntary agricultural conservation programs—in this case, to reduce farming emissions and increase carbon capture in crops, trees, and soil.

Some of these programs would have been infeasible if not for the groundwork laid by private industry over the prior decade. One of the most remarkable advances is evident in the desert just a few miles east of my clinic. There, in 2016, Tesla, in partnership with Panasonic, opened an enormous gigafactory to produce batteries for its cars. Many of my patients' parents now work in that factory, which runs twenty-four hours a day to try to meet demand.

In response to Tesla's success, all the major auto manufacturers—including GM, Ford, and Volkswagen—have announced plans to electrify their fleets in coming decades, and many more affordable models are in the works. Meanwhile, more than a dozen states, including Nevada, have adopted "clean car" regulations to promote electric vehicles, and the US Environmental Protection Agency recently made a similar move. So while these cars were expected to account for only 16 percent of global auto sales in 2023, that share is six times what it was in 2019, and guaranteed to grow. And Tesla's rapid expansion since 2016 should give us hope for what's possible in the decade ahead.

At the same time, many local utilities now allow customers to buy some or all of their electricity from green sources. Enrolling takes just a few clicks of a mouse and doesn't require doing anything structural to a home, but often comes with a small surcharge (at my house in Nevada, it's less than $1.50 a month). Some Americans have already eliminated over half their greenhouse gas emissions by driving electric cars and plugging them into homes powered by green energy, either from their utility or rooftop solar panels. Though these steps have been financially out of reach for most people, incentives in the IRA will make them much more affordable.

There are many other actions parents can take; they can be found in books like *Drawdown*, edited by Paul Hawken, and its accompanying website, drawdown.org. Climate-helping projects for children can be found in sources like *The Parents' Guide to Climate Revolution*, by Mary DeMocker. Some climate activists object to these types of activities, arguing that any focus on individual behavior is inappropriate. Instead, they say, we should push for a greater corporate reckoning. They rightly point out that only one hundred companies have produced nearly three-quarters of all the greenhouse gasses emitted since NASA scientist James Hansen warned Congress about climate change in 1988—and that the fossil fuel industry has intentionally promoted the idea of a personal "carbon footprint" to distract from its own culpability.

But we don't need to choose. The actions of individuals, public institutions, and corporations are not competitive, they're iterative: they feed off one another. There would be no Tesla if individual people had not bought its cars—and without Tesla, the rest of the automotive industry would not have followed. Utilities would not have expanded their green energy programs if customers never chose that option. And, as the IRA underscores, our elected government always plays a role in kick-starting and shaping industry: Tesla received federal and state assistance to build the electric car; utilities can provide renewable energy only because of regulatory changes that allowed it.

Individual action is also important for how it affects our minds. Action breaks us out of despair and inertia. It gives us a sense of agency over this global crisis that we might otherwise surrender to. And that sense of power and hope can be contagious; it can turn a spoon against the ocean into a bucket, and then a brigade.

Fighting Back

No group has demonstrated the power of individuals better than the youth climate activists. As they gathered by the thousands at the twenty-sixth UN

Conference of the Parties (COP26) in Glasgow, Scotland, in 2021, many in the crowd noted that their movement had started with a single teenage girl in Sweden, sitting in front of her parliament once a week with a handmade sign. Greta Thunberg's Fridays for Future, and other youth groups such as Sunrise Movement, have shown that the "climate grief" I've seen in some of my patients can be lifted with protest. Anger, like hope, is an energizing emotion. Children and teens may lack money or the vote, but they have found moral power by fighting the injustice being done to them.

Young people have also taken their protest to the courts, most famously in *Juliana v. United States*, in which a group of child plaintiffs argued that the federal government's ongoing policies allowing climate change must stop because they have violated the youngest generation's "fundamental constitutional rights to life, liberty, and property." Similar lawsuits have been filed at state and international levels, but until recently these efforts have been stymied by the courts. In *Juliana*, for example, the Ninth Circuit ruled that the children—incredibly—had no legal standing for their claims.

A breakthrough came in August 2023, however, when a Montana judge ruled in favor of a group of young people who had sued the state over its failure to consider climate change when reviewing permits for fossil fuel projects. The court found that the state had violated the children's "right to a clean and healthful environment," as guaranteed by the state constitution. Some of the children testified that worsening air pollution, wildfires, and heat had harmed their health, and a local pediatrician, Dr. Lori Byron, was a star witness for their case. Though the decision is being appealed, activists hope that *Held v. Montana* marks a legal turning point.

That same month, young and indigenous activists in Ecuador achieved another remarkable feat. As a result of their decade-long campaign, Ecuador's citizens voted for a referendum that bans new oil extraction in Yasuní National Park, a part of the Amazon considered to be one of the most biodiverse places on Earth and home to some of the last uncontacted indigenous peoples anywhere. Though its existing oil extraction projects will continue, Ecuador—a small, impoverished country—was the first in the world to vote to keep oil in the ground.

The annual COP meetings have been less encouraging. Despite the youth groups flexing their muscle, Glasgow itself was a mixed bag, and many activists expressed deep disappointment in its results. While they chanted in the streets, five hundred representatives of oil, gas, and coal interests—more than the delegation of any single country—were scheming quietly in the halls, sabotaging the demands of the generation who will inherit their handiwork.

By the time COP28 was held in Dubai in 2023, the number of fossil fuel lobbyists had swelled to nearly 2,500. The conference was held in one of the world's leading oil-producing countries and chaired by a sultan who runs an oil company. The resulting agreement was hailed for finally naming fossil fuels as the main cause of the crisis—the first time in three decades that had happened. But its vague call for an eventual transition away from those fuels was devastating to anyone who understands the urgency of the moment. A World War II–type mobilization is needed to cut emissions in half this decade. Yet eight years after the Paris Agreement at COP21—when the world committed to try to limit warming to 1.5 degrees—the fossil fuel industry has corrupted the process for meeting that goal.

Climate scientist Michael Mann and writer Susan Joy Hassol have noted the "understandable anxiety, despair, and righteous anger on the part of young people given the insufficiency of the progress and the bad actors who are creating obstacles." As the crisis deepens, these teens and young adults are watching powerful people try to bargain with reality. Many still don't seem to grasp that the atmosphere doesn't care about elections or shareholders, and the laws of physics don't bend.

"Dear Friends"

Some children have raised their voices in other ways.

Lillian Fortuna was only ten years old when she first read Greta Thunberg's book *No One Is Too Small to Make a Difference.* "It was like a wave had hit, like it was all really, really real," she tells me two years later. "It made me kind of hate humans, and what they've done to the planet. I got really sad about that, and angry."

That was in the spring of 2020, during the first, terrible COVID-19 surge in New York City. She and her family had temporarily moved from their Brooklyn apartment to a remote farmhouse upstate—where, stuck in isolation, the small fourth grader began taking on the weight of the world. "I felt like I was just sitting here, reading about climate change and watching it happen, and there was nothing I could do."

Now twelve years old and in the seventh grade, Lillian is back in Brooklyn, speaking to me over Zoom with her mother by her side. She's a young gymnast and looks it: petite and fit in a loose blue T-shirt, her straight red hair pinned up in back. "She went to a really dark place," her mom, Mila, explains. Mila told her daughter that a lot of people, adults and kids, feel the same way about the climate crisis: they want to help but don't know how. She suggested Lillian write a letter a month to friends and family, each dedicated to just one thing people could do. "Action seemed like it would be the best antidote to her despair," Mila says, "and Lillian needed to see that people cared."

After some research, Lillian decided to write her first letter not about global warming, but a related passion: the plight of animals. "Dear Friends," she began, before describing the mistreatment of monkeys forced to pick coconuts in Southeast Asia. "The monkeys do have strong feelings about this," she said, reporting that one of them had killed an overseer with a well-thrown coconut. She urged her readers to boycott companies that exploit monkey labor. Then she signed the letter "Sincerely, Lillian," and sent it.

When the letter got a positive reaction, her spirits lifted. She has since composed over seventy "Dear Friends" letters, each about a different animal or environmental topic, many of them focused on climate change. And she's set up a website, madplanet.org, where she posts them for a wider audience.

But her most influential project came in response to another global calamity: Russia's invasion of Ukraine in early 2022. Just three days after the attack began, Lillian read an article by climate activist Bill McKibben. In it, he argued that President Biden could address both the climate and Ukraine crises in one step: by invoking the Defense Production Act (DPA) to produce millions of heat pumps for Europe. Heat pumps are highly efficient, electric alternatives to furnaces and boilers; they can run on renewable energy like solar and wind. A rapid European transition to heat pumps would bring a hat trick of benefits: it would reduce the continent's reliance on Russian oil and gas, slash the Russian military's main source of funding, and cut greenhouse gas emissions that endanger everyone on the planet.

Lillian was inspired. She decided to start a petition on change.org, urging the president to follow McKibben's advice. Along with writer Rebecca Solnit, McKibben got wind of the petition, titled "Stop Putin by Sending Heat Pumps to Europe," and signed it; he later interviewed Lillian in a brief video that he posted online. Within weeks her petition had topped three thousand signatures and made it to the president's desk.

"It was just unbelievable that Rebecca Solnit and Bill McKibben both reached out to her," her mom says. "The impact of these heroes of hers saying: 'I see you, let's do this together, what's our next thing?' was profound."

An even more exciting event soon followed. In June 2022 Biden did, in fact, invoke the DPA to boost manufacturing of heat pumps and other green technology, like solar panels. In October, Lillian received a personal letter from the president. "It's important that you continue to use your voice to speak out on this issue," he wrote. "Always remember, when you make your voice heard, adults listen."

"That was really cool," she says, and smiles.

After hearing how much McKibben and Solnit's support had meant to Lillian, I reached out to the two writers to ask the reverse: how did this young girl's petition and activism affect *them*? "I think it made everyone happy," McKibben answered, "in part because she's so wonderfully serious." He had spoken to her through his work with Third Act, an organization he founded to get over-sixties involved in the climate fight. Much of their activism, he says, is "across the age divide," working with young people like Lillian.

Solnit took a broader view. "I both feel terrible that kids have to think about these big menacing things and excited that they find their way to take action," she responded, "both because that action matters in the larger picture and because it's such a great cure for feeling powerless and helpless." In that light, Solnit said, she knew Lillian's example was inspiring for a lot of other people—including herself.

Lately, Lillian's advocacy has aimed closer to home. After the success of the Ukraine petition, her mom convinced the board of their fifty-five-unit Brooklyn co-op to pause on replacing the building's aging boiler and investigate whether they could install heat pumps instead. "Lots of New York buildings are in the exploratory phase on this," Mila says, referring to New York City's plan to decarbonize its buildings. "It's a huge undertaking, but if we can figure it out, others will follow." She looks over at her daughter with respect and affection. "Of course, I have to try." The board also plans to apply a reflective coating to the building's roof to reduce heat absorption in summer, and install a water heater that's powered by a heat pump. It is looking into solar panels, as well.

"It's really cool to learn about these things and then see people take action," Lillian says, and grins. "Even if it's my parents." Her dad has also gotten in on the act, trying to persuade the company he works for to divest from banks that fund fossil fuel projects. That's her latest passion: how financial institutions use our money to promote fossil fuels, and why we should move our accounts to banks that don't. Of course, she's planning on writing more letters.

❁

Lillian is a delightful kid, very smart and a joy to talk to, earnest but also funny. She laughs at herself as she tells me about going to the 2019 Climate Strike in New York City with her mom and demanding to get lemonade instead of staying to hear Greta Thunberg speak—because at the time, she didn't know who she was.

Now, like the famous activists she admires, this one young person has set off ripples of change that will hopefully spread, meet, and grow. Did she have any advice for those who want to follow her example? "All you have to do is learn about climate change and tell others about it. That does a lot itself." She adds that it's important to stay positive. "I try to make sure people understand there are ways to help and it's not all over."

Before ending our call, I ask Lillian what she would say to the heads of ExxonMobil or Chevron if she had the chance. "I'd try to make them aware of the harm they're causing and the actual amount of time we have left, because it's really alarming. If they don't understand that, it might be one reason why they're still doing what they're doing."

It was such a reasonable and innocent thing to say. Why would anyone with power, who understood climate change as Lillian does, *not* act to protect her and other young people, as well as all the animals and nature she loves?

Fortunately, most adults are not as callous as oil company executives, and children like Lillian are not alone in this struggle. Those who care about them are battling, too. Because fighting back isn't just the best way to cope. It's the only way to win.

Every Pound of Carbon

Climate activists proved this repeatedly in 2022, when years of hard work finally yielded several major victories. The IRA was the most significant. But just a month after its passage, the US Senate also ratified the Kigali

Amendment, an international treaty to curtail production of hydrofluorocarbons (HFCs). Used mainly in refrigerators and air conditioners, HFCs are powerful greenhouse gases. The treaty will lead to their replacement with new, climate-friendly alternatives, preventing up to 0.5 degrees Celsius of warming by 2100.

A handful of states also broke new ground against carbon dioxide and methane pollution, with California leading the charge. Home to one of the largest economies in the world, the state announced that it will soon ban the sale of fossil fuel–powered cars, furnaces, and hot water heaters, a decision sure to reverberate through the country.

Much like when the US joined World War II, these moves have shifted global momentum. "After so many years of doing next to nothing on the climate," *The Atlantic* gushed, "America seems like it's in the middle of a mini–golden age for climate policy making." But as was the case in 1941, the years ahead are difficult, malevolent forces are still powerful and battling, and the outcome is far from certain. What happens next will depend on all of us.

Because while humanity has the tools to hold Earth's warming to 1.5 degrees Celsius, we are currently on track to shoot past that goal. Current international pledges, if kept, are expected to cause warming of 2.5 to 2.9 degrees above preindustrial times by 2100. While that's an improvement over projections from just a few years ago, it still portends a terrifying and unacceptable future. "People do not understand the magnitude of what is going on," climate scientist Katharine Hayhoe has said. "This will be greater than anything we have ever seen in the past. This will be unprecedented. Every living thing will be affected."

This is the *other* world that children like Anna could inherit. A world engineered by powerful people who took their profit and unloaded the cost of their business onto everyone else. One in which many arable regions will be transformed into uninhabitable, barren land, triggering mass starvation, violence, and migration. Where many ecosystems and species will be lost. Where sea levels will rise too fast for coastal cities to move inland and adapt. Where many children will be hurt, their potential cut short.

We already live in the beginnings of that world. We can't do anything about the warming that's happened, disrupting our environment and society even now. We can see its impact on human health; I see it in my clinic.

But each pound of carbon, each tenth of a degree, and each month of delay will make a difference in the world we leave our children. We must refuse to cooperate with the destruction of their future. We can use our voice, ballot, and wallet to fight for clean energy and a rapid end to oil, gas, and coal. Children are facing the equivalent of a medical emergency, and the fossil fuel industry is blocking the ambulance. Treatment that comes too late cannot help. Prevention is the best medicine of all.

So when parents ask me what they can do, I say: start by looking at your house, your apartment, your neighborhood, your workplace, your city and state, and asking questions. What gas-powered appliances do you have in your home? Could you replace them with electric alternatives, with help from government programs? When you recharge your phone or heat food in the microwave, where does the electricity come from? Where is the local power plant, and what is it burning? Does your utility have a plan to end its use of fossil fuels and convert to renewable sources? How does the electricity get from the power plant to you? Does the grid meet your community's needs?

Utilities, city councils, and state legislatures hold meetings on energy-related and urban planning topics all the time, and public input is often sought. You can speak in support of clean energy and green spaces and explain what they mean for your children's health. You can also learn how much carbon your city or state emits each year, and where it comes from: transportation, power plants, industry, garbage dumps, buildings, agriculture. What is your local or state government doing to reduce emissions from each of these major categories?

Many states and municipalities have prepared "climate action plans" and greenhouse gas inventories that summarize this information. If your state hasn't, you can lobby for it to do so. If it has, look for a graph or table that compares current emissions and their sources to goals for 2030 and 2050. Is your community on track to meet its goals? If not, what more needs to be done?

You and other parents can use these goals to press public agencies and others in the community to move in the right direction. Does your electric company offer a renewable option, and if not, why not? Has your landlord or boss signed up for that option, or provided an outlet in your parking garage for car charging? Will your city council or state legislature take steps to reach "net zero" quickly, by, for example, banning natural gas in new home construction? And what are they doing to prepare for rising heat, wildfires, drought, flooding, and storms, especially in low-income neighborhoods that are often most vulnerable to disaster?

If you are fortunate enough to have money saved or invested, you should look closely at how it is being used. The country's largest banks—Chase, Citibank, Wells Fargo, and Bank of America—continue to fund fossil fuel projects; if you have a credit card or savings account with one of them, it may be the biggest contributor to your individual carbon footprint. Has your company's retirement fund or your state's pension fund stopped investing in fossil fuels—or cryptocurrency, which requires staggering amounts of energy usually generated by those fuels? How about the endowment fund for the college you attended?

Your child's school is a great place both to tackle climate change and improve kids' health. You can push the school to replace furnaces and air conditioners with heat pumps and solar panels; your child might want to join in (or lead) the effort. A nationwide movement to decarbonize schools and create more natural school playgrounds is growing. Multiple states are transitioning from diesel school buses to electric models, and new federal funding provides $5 billion to help school districts make this switch. Two nonprofit groups, the New Buildings Institute and the World Resources Institute, have produced roadmaps for school districts wanting to take these steps, which also reduce children's exposure to air pollution and save schools money in the long run.

Our jobs provide other opportunities. Regardless of what you do for a living, you can join with others to advance climate-friendly practices and policies in your work. For instance, I belong to a group of climate-concerned pediatricians who meet on Zoom once a month to share research, patient education materials, and news about climate legislation and advocacy from every state. Another group of physicians has formed Health Care Without Harm, which works to reduce greenhouse gases and waste from hospitals and clinics. Others are writing op-eds, testifying at public hearings, and speaking at conferences to bring more colleagues into the fight.

Plant a tree, or save one from being cut down. Vote for leaders who will end fossil fuel subsidies—which globally total trillions of dollars a year and distort the market in favor of oil and gas. Stop buying products containing palm oil, a key driver of rainforest destruction. Campaign for green building codes in your state. Try to reduce your use of plastics and other petrochemicals; think about what your habits and purchases really cost. Talk to others about what you know. "We can't give in to despair," Katharine Hayhoe says. "We have to go out and look for the hope we need to inspire us to act—and that hope begins with a conversation, today."

But if you do despair—if you hear the lie that it's too late, too expensive, or too difficult to save your children from the worst; or that solar panels, wind turbines, and electric cars are as bad for the environment as fossil fuels; or that people will freeze or starve if we give up those fuels, when in fact continuing to burn them threatens billions of lives—remember the hidden hand behind many of these stories. As Michael Mann detailed in *The New Climate War*, fossil fuel corporations and their allies, including authoritarian petrostates such as Saudi Arabia and Russia, want us to feel hopeless, to give up and disengage. Through bots, trolls, and propaganda, they push for our surrender to the fate they've chosen. We must take the wheel from them and turn.

Finally, I hope we can all do something small but meaningful. When we pack our kids' lunches, help them with homework, or kiss them on the forehead, I hope we can notice the air around us. Take a deep breath, pull it into our cells. Pay attention to how it feels. Have our children do the same. Unplug them and go outside. Reconnect them to their bodies and to physical reality. Teach them—not with words, but with experiences—to be mindful of their place in the tapestry. Because they are part of the atmosphere and all its life. And every breath they take weaves them to this unique and precious planet that is their home.

Ashes to Flowers

Not long ago I received a video from the mother of one of my patients, a five-year-old boy. He had jogged up a flight of stairs without using the rail; she wanted me to see it.

That might not seem like a remarkable feat. But for his parents it was a miracle. He has Duchenne muscular dystrophy. A year before the video was shot, he could hardly climb stairs at all.

Duchenne is a genetic illness that progressively weakens the muscles. In the past, affected boys rarely lived beyond young adulthood. But in 2021 my patient was given an experimental drug that researchers hope will correct his faulty gene. Five weeks later, this child who'd been unable to rise easily from the floor was beating his sister in a race up the stairs.

I watched the video knowing that his small but momentous act was the end point of countless threads through history. Of so much work, by so many people—parents, doctors, scientists, and others—who had come together in common purpose. Supported by a democratic government that had funded this advance and the infrastructure necessary to deliver it.

And I knew that I was part of those threads, in more ways than one. After texting his mother back and sharing in her joy, I looked at a round, pocked scar that I've carried on my arm since kindergarten. It's my personal remnant of humanity's battle against smallpox, "the most terrible of all the ministers of death." The disease killed an estimated 300 million people in the twentieth century alone before being eradicated by a vaccine. I was among the last groups of schoolchildren to receive it.

We are linked to each other in a chain of generations, responsible to both past and future. The fight against climate change is not just for that little boy, or Anna, or all the other children I know. It is also for their parents and grandparents and great-grandparents. It is what we owe to the struggles and accomplishments of those who came before us, who overcame other crises that felt insurmountable. We can't allow our inheritance to be lost.

To the parents who crossed deserts and mountains and borders to get a better life for their children. To the scientists who built satellites to save us from hurricanes, and warn us that we were changing the Earth.

To the doctors and researchers who developed vaccines and medicines that have saved millions of lives.

To all of them, we owe the possible world.

When I met Shani and her family in Santa Rosa after the Tubbs Fire, I sat on a park bench as she and her brother played. She asked me and her mother to watch her do cartwheels in the grass. After a couple of tumbles she stopped, plucked two dandelions from near her feet, and brought one to each of us. "Do you want to see my drawing?" she asked, leaning against my thigh.

"Of course," I said.

She asked for her mother's purse and pulled it out. Along the entire bottom of the page was a row of frightening orange flames. Above that was a thick layer of black and gray, in chaotic crayon strokes. And above that, flowers, green grass, and sunshine.

"This is what happened to us, the fire," she said, pointing at the flames. Then she moved her finger up to the dark, angry layer. "This is the ash, and how it felt when everything was gone. But this shows what the future will be." She pointed to the bright upper part of the drawing.

None of us knows her future. Yet we are part of it and everything that follows; part of whether ash becomes life.

I went back to my hometown later that year, and stood at the edge of the marsh. At my old elementary school a block away, a half-staffed American flag whipped in the wind, its pole rings clanging. Children were walking by, headed to the same houses around the marsh that we had once called home. A little girl with a knapsack as big as her body kept trying to balance-beam the curb and getting gusted off by the wind. She had a sticker planted off-kilter on her chest: *The sky's the limit!*

She looked at me for a moment as she passed. And I watched her round the corner, onto the street where I had lived when I was young.

In the small, one-story house she would soon see to her left, I had fallen asleep beneath the feet of my father's recliner as we watched men first walk on the moon—less than a decade after America had set that goal. Everyone on Earth that night had realized, together, that what seemed impossible could be done.

I hoped the girl would come back to the marsh and look for water skeeters and crabs and polliwogs. I hoped that, like me, she felt a part of this place. Then I looked down at a blade of marsh grass in my hand, picked from the ground a few minutes before. It was dusted with ash from Paradise, the small city that had recently burned down.

She won't know the world I knew. I brushed my hand across the tops of the reeds. Bright green erupted from beneath the gray powder, which rose in a small cloud and was swept away, in an instant, by the wind.

A great blue heron was fishing nearby, in the creek that runs through the marsh. He saw me watching him, tipped his head, and eyed me as if to scold, *It's still here to fight for.*

I got in my car and started driving east across California toward my home, on the other side of Donner Pass. As I crested it a few hours later, the sky over Nevada, stretching out in front of me, was blue and clear and vast.

Surely our voices produce more than sound. Surely our science will yield more than a chronicle of our end.

Because climate change has a face, and it is a child's.

Acknowledgments

I owe thanks to many dear friends, family members, and colleagues who supported, encouraged, and listened through my years of researching and writing this book. At the top of this list is my friend Patrick Wright, director of the Governor's Wildfire and Forest Resilience Task Force in California. He was my first reader and the person who has reviewed the manuscript through its many iterations more than anyone else. Without his initial reaction, coaching, criticism, editing, prodding, and expertise, the book would not exist.

Deep thanks are also owed to my friend Barbara Kohlenberg, who suggested I show the first pages to Alan Deutschman, a writer and journalism professor at the University of Nevada in Reno. I hoped at the time to publish them as a magazine essay, but I am forever grateful to Alan—who would become a friend and mentor over many visits to the local coffee shop—for suggesting that we send them to author Katherine Boo, who had just visited the university. I cannot thank Kate enough for then forwarding those pages to her agent, Amanda "Binky" Urban, to whom I am also indebted. It was Binky who asked me to write more, helped me shape my ideas into a book, signed me on, and navigated the project through to the end. I want to thank my editors at Simon & Schuster: Priscilla Painton for her patience and guiding hand, and Hana Park for her doggedness and passion in getting the book done. I also owe special thanks to production editor Morgan Hart and the rest of her team for their meticulous attention to detail and flexibility regarding questions of voice and style.

I give thanks and love to my sisters, Stacy Hendrickson and Debra Granato; my wonderful friends Elizabeth Gifford, Jennifer Waltz, Steven Graybar, Lyn Stein, Dosheen Cook, and Carole Greenfield; my aunt Sharon Knowles; and Robert Levy, who has been a second father to me since my own father's passing. All of them surely got tired of hearing about this topic, but showed me nothing but unwavering support.

I need to thank my longtime business partners, colleagues, and friends, Sheryl Cohen, Kathleen Christopherson, and Joel Speicher, as well as our new partners, Jennifer Ehmann and Jessica Vaughan, for their flexibility, understanding, and support as I wrote a book while working as a physician in a private practice. And our wonderful staff, who helped me manage many scheduling challenges and patient care needs.

I am grateful to Samantha Ahdoot, a Virginia pediatrician, founding member of Virginia Clinicians for Climate Action, author of the American Academy of Pediatrics' climate change policy statement, and friend, who has been endlessly gracious and generous in her support of me and the book; to Lori Byron, a Montana pediatrician and head of the American Academy of Pediatrics' Chapter Climate Advocates network, for her tireless leadership and encouragement; and to Leann DiDomenico McAllister, executive director of the Nevada Chapter of the American Academy of Pediatrics, and Joanne Leovy, head of Nevada Clinicians for Climate Action, for their support of the book and climate advocacy in our state. I also thank Mary Data, my research assistant, for her many trips to the medical school library and help with the ever-growing list of endnotes.

I am indebted to the scientists and researchers who are named in these pages: Jennifer Vanos, David Hondula, Nancy Selover, Julie Kaplow, and Amy Vittor. All were very giving of their time and expertise, and in some cases read draft chapters to be sure I'd gotten the science right. I also appreciate the information shared with me by Daniel Kiser and William "Jim" Metcalf of the Desert Research Institute; Vjollca Berisha and her heat illness team at Maricopa County Public Health; Brent Kaziny, medical director of emergency management at Texas Children's Hospital; Michael Braun, Helen Currier, and staff of the Renal Center at Texas

Children's Hospital; David McClendon and Anna Hardway of Save the Children; Ivan Gonzalez and Christine Curry of the University of Miami Miller School of Medicine; Dean Blumberg of the University of California Davis Children's Hospital; Hao Tran, the Iowa pediatrician who cared for Camron; and Melissa Ramos of the American Lung Association office in Las Vegas. I am particularly grateful to the Phoenix paramedics who shared, on condition of anonymity (due to legal issues surrounding one of the cases), the medical details and personal toll of responding to heat rescues, especially those involving children. And I deeply appreciated the conversation I had with Keiko Ogura, who survived the bombing of Hiroshima at the age of eight, about the impacts of catastrophe on children.

Most of all, of course, I owe thanks to the families who agreed to be interviewed or portrayed in these pages, including several of my own patients and their parents. For some, the experiences they shared were the worst thing that had ever happened to them. In all cases they found the topic difficult but provided timelines, details, and photographs to try to help. I am honored and moved that they trusted me to tell their children's stories. And I am grateful for getting to know such caring and good people through this process.

Finally, I want to thank my own parents, Charles and Sandra Hendrickson, for raising me in beautiful and wild places where I felt part of the fabric of living things. And I thank my three now adult children. When they were small, they were the reason I became interested in pediatrics, and our love for one another is still the best thing in my life. For them and their children, I hope and work for the possible world.

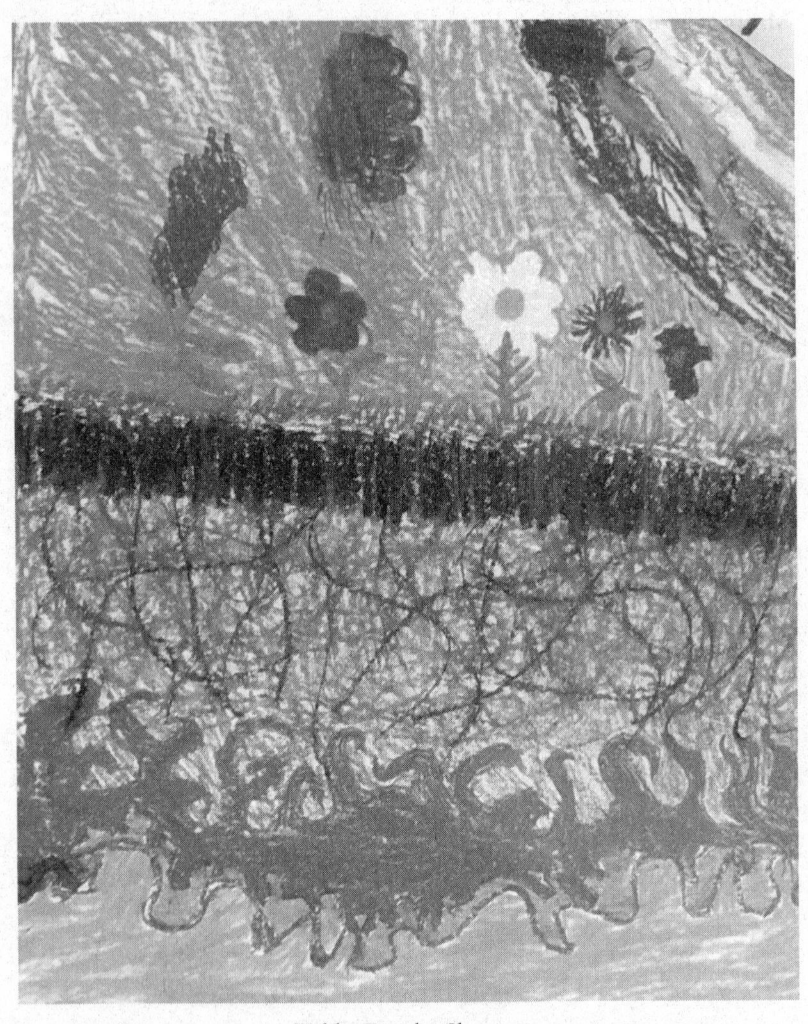

Tubbs Fire, by Shani, age 5

Endnotes

Epigraph

ix *Without showing what happened to a child*: Keiko Ogura, personal communication, Hiroshima Peace Memorial Museum, Hiroshima, Japan, August 1, 2017.

Introduction: Heroes

2 *Global temperatures rise at an alarming pace*: National Aeronautics and Space Administration (NASA), "Global Temperature," *climate.nasa.gov*, accessed February 26, 2023, https://climate.nasa.gov/vital-signs/global-temperature/.

2 *The eight years prior to 2023 were the hottest ever recorded*: Henry Fountain and Mira Rojanasakul, "The Last 8 Years Were the Hottest on Record," *New York Times*, January 10, 2023, accessed March 10, 2023, https://www.nytimes.com/interactive/2023/climate/earth-hottest-years.html.

3 *Then 2023 shattered Earth's heat records*: Zeke Hausfather, "State of the Climate: 2023 Smashes Records for Surface Temperature and Ocean Heat," *carbonbrief.org*, January 12, 2024, accessed January 15, 2024, https://www.carbonbrief.org/state-of-the-climate-2023-smashes-records-for-surface-temperature-and-ocean-heat/.

3 *This relentless warming trend would quickly flatten*: Mark Hertsgaard, Saleemul Huq, and Michael E. Mann, "How a Little-Discussed Revision of Climate Science Could Help Avert Doom," *Washington Post*, February 23, 2022, accessed March 23, 2023, https://www.washingtonpost.com/outlook/2022/02/23/warming-timeline-carbon-budget-climate-science/.

3 *Surveys show that a large majority of children*: Caroline Hickman, Elizabeth Marks, Panu Pihkala et al., "Climate Anxiety in Children and Young People and Their Beliefs About Government Responses to Climate Change: A Global Survey," *Lancet Planetary Health* 5: 12 (2021): e863–e873; Mathew Ballew, Jennifer Marlon, Seth Rosenthal et al., "Do Younger Generations Care More About Global Warming?," *climatecommunication.yale.edu*, June 11, 2019, accessed March 10, 2023, https://climatecommunication.yale.edu/publications/do-younger-generations-care-more-about-global-warming/.

4 *I am a pediatrician in the fastest-warming city*: Climate Central, "Earth Day: US Warming Rankings," *climatecentral.org*, April 20, 2022, accessed February 26, 2023, https://www.climatecentral.org/climate-matters/earth-day-warming-rankings.

4 *The American Academy of Pediatrics has warned*: Katherine M. Shea and Committee on Environmental Health, "Global Climate Change and Children's Health," *Pediatrics 120*: 5 (2007): e1359–e1367; Samantha Ahdoot, Susan E. Pacheco, Council on Environmental Health et al., "Global Climate Change and Children's Health," *Pediatrics 136*: 5 (2015): e1468–e1484; Ying Zhang, Peng Bi, and Janet E. Hiller, "Climate Change and Disability-Adjusted Life Years," *Journal of Environmental Health 70*: 3 (2007): 32–38.

6 *In 1977, when I was about this girl's age*: James F. Black, "The Greenhouse Effect," presentation to Exxon Corporation Management Committee in July 1977, updated version presented to Exxon research and engineering on May 18, 1978, transcript posted by *insideclimatenews.org*, accessed July 5, 2023, https://insideclimatenews.org/wp-content/uploads/2015/09/James-Black-1977-Presentation.pdf.

6 *But by 1988, when NASA scientist James Hansen*: Philip Shabecoff, "Global Warming Has Begun, Expert Tells Senate," *New York Times*, June 24, 1988, accessed July 5, 2023, https://www.nytimes.com/1988/06/24/us/global-warming-has-begun-expert-tells-senate.html; Geoffrey Supran and Naomi Oreskes, "Assessing ExxonMobil's Climate Change Communications (1977–2014)," *Environmental Research Letters 12*: 8 (2017): 084019; Neela Banerjee, John H. Cushman, David Hasemyer, and Lisa Song, *Exxon: The Road Not Taken* (Inside Climate News, 2015); Center for Climate Integrity, "Documentary Evidence of Oil and Gas Companies' Knowledge of Their Products' Role in Causing Climate Change and Their Subsequent Deception Campaign," *climateintegrity.org*, accessed July 5, 2023, https://climateintegrity.org/uploads/media/DeceptionBinder_September2019.pdf.

6 *Outside the windows of the medical center*: Benjamin Spillman, "Reno Just Baked Through Hottest Month on Record—Get Used to It," *Reno Gazette Journal*, August 1, 2018, accessed March 10, 2023, https://www.rgj.com/story/life/outdoors/2018/08/01/reno-just-baked-through-hottest-month-ever-get-used/876965002/.

6 *In the Sierras and beyond, heat was baking California's forests*: US Forest Service (USFS) and California Department of Forestry and Fire Protection (CalFire), "Record 129 Million Trees Dead in California," *fs.usda.gov*, December 12, 2017, accessed March 3, 2023, https://www.fs.usda.gov/Internet/FSE_DOCUMENTS/fseprd566303.pdf.

7 *At night, it would trouble their sleep*: Rebecca Lindsey, "Extreme Overnight Heat in California and the Great Basin in July 2018," *climate.gov*, August 8, 2018, accessed July 2, 2023, https://www.climate.gov/news-features/event-tracker/extreme-overnight-heat-california-and-great-basin-july-2018.

11 *Because we can still save our children*: Valérie Masson-Delmotte, Panmao Zhai, Hans-Otto Pörtner et al., eds., Intergovernmental Panel on Climate Change (IPCC), *Global Warming of 1.5°C: An IPCC Special Report on the Impacts of Global Warming of 1.5°C Above Pre-industrial Levels in the Context of Strengthening Response to Climate Change, Sustainable Development, and Efforts to Eradicate Poverty* (Cambridge, UK: Cambridge University Press, 2022), accessed July 5, 2023, https://www.ipcc.ch/site/assets/uploads/sites/2/2022/06/SR15_Full_Report_LR.pdf.

11 *James Black's granddaughter recently observed*: Nermeen Shaikh, "Granddaughter of Exxon Scientist Confronts CEO over Oil Giant's Decision to Fund Climate Lies,"

democracynow.org, May 26, 2016, accessed March 5, 2023, https://www.democracy now.org/2016/5/26/granddaughter_of_exxon_scientist_confronts_ceo.

12 *"If everyone is guilty"*: Ivana Kottasova and Eliza Mackintosh, "Teen Activist Tells Davos Elite They're to Blame for Climate Crisis," *CNN*, January 25, 2019, accessed March 5, 2023, https://www.cnn.com/2019/01/25/europe/greta-thunberg-davos -world-economic-forum-intl/index.html.

12 *Later that year, tens of thousands of children*: Somini Sengupta, "Protesting Climate Change, Young People Take to Streets in a Global Strike," *New York Times*, September 20, 2019, accessed March 5, 2023, https://www.nytimes.com/2019/09/20 /climate/global-climate-strike.html.

Chapter One: Breathing

16 Avoid spreading ash: Michael Lipsett, Barbara Materna, Susan Lyon Stone et al., *Wildfire Smoke—A Guide for Public Health Officials* (Sacramento, CA: Office of Environmental Health Hazard Assessment, July 2008), appendix C, p. 41.

16 *Two weeks earlier, a careless hunter*: Kate Mather, "Hunter Indicted in Huge Rim Fire Near Yosemite," *Los Angeles Times*, August 7, 2014, accessed July 7, 2023, https:// www.latimes.com/local/la-me-rim-fire-20140808-story.html; CalFire, "Top 20 Largest California Wildfires," *fire.ca.gov*, October 24, 2022, accessed March 12, 2023, https://www.fire.ca.gov/our-impact/statistics.

16 *Day after day, a relentless conveyor belt of wind*: US Census Bureau, "QuickFacts: Reno City, Nevada; Spanish Springs CDP, Nevada; Sparks City, Nevada: Population, Census, April 1, 2010, and Population Estimates, July 1, 2022," accessed March 12, 2023, https://www.census.gov/quickfacts/fact/table/renocitynevada,spanishsprings cdpnevada,sparkscitynevada/POP010210.

16 *Even for those with normal lungs*: Jeff DeLong, "Smoke from Yosemite Fire Brings Unhealthy Air to Area," *Reno Gazette Journal*, August 27, 2013, accessed March 5, 2023, https://www.usatoday.com/story/news/nation/2013/08/27/yosemite-fire -smoke-air-quality/2705527/.

17 *Since Anna's visit to my clinic*: CalFire, "Statistics," *fire.ca.gov*, accessed March 12, 2023, https://www.fire.ca.gov/our-impact/statistics.

Donner Pass

19 *They should have crossed before the snows*: Frank Mullen Jr., Marilyn Newton, Will Bagley, and Rollan Melton, *The Donner Party Chronicles: A Day-by-Day Account of a Doomed Wagon Train, 1846–1847* (Reno and Las Vegas, NV: A Halcyon Imprint of the Nevada Humanities Committee, 1997).

Mind the Atmosphere

21 *The atmosphere is only as thick*: NASA, "The Atmosphere," *grc.nasa.gov*, accessed July 6, 2023, https://www.grc.nasa.gov/www/k-12/airplane/atmosphere.html.

22 *It's no wonder fossil fuel companies*: Michael E. Mann, "Chapter 2: The Climate Wars," in *The New Climate War: The Fight to Take Back Our Planet* (New York: Public Affairs, 2021), pp. 21–45; Center for Climate Integrity, "Documentary Evidence of Oil and Gas Companies' Knowledge of Their Products' Role in Causing Climate Change and Their Subsequent Deception Campaign."

22 *We understand well the mechanics of breathing*: US Department of Commerce, National Oceanic and Atmospheric Administration (NOAA), "Atmosphere: Introduction to the Atmosphere," *noaa.gov*, last modified April 23, 2023, accessed June 15, 2023, https://www.noaa.gov/jetstream/atmosphere.

The Best of Times, the Worst of Times

23 *In 1853 more than half of all deaths*: US House of Representatives, "Introductory Chapter," *Mortality Statistics of the Seventh Census of the United States, 1850* (Washington, DC: 33d Congress, 2d Session, Ex. Doc No. 98, 1855), p. 16, accessed July 6, 2023, https://www2.census.gov/library/publications/decennial/1850/1850b/1850b-02.pdf.

23 *"My poor boy"*: Elizabeth Keckley, *Behind the Scenes, or Thirty Years a Slave, and Four Years in the White House* (New York: G. W. Carleton & Company, 1868), p. 103.

23 *Willie's physician, Dr. Robert Stone*: Jean H. Baker, *Mary Todd Lincoln: A Biography* (New York: W. W. Norton & Co., 1987), p. 209.

23 *Through vaccines, water treatment plants, antibiotics*: Sherry L. Murphy, Kenneth D. Kochanek, Jiaquan Xu, and Elizabeth Arias, "Mortality in the United States, 2020, NCHS Data Brief No. 427," National Center for Health Statistics (2021), *cdc.gov/nchs*, accessed March 10, 2023, https://www.cdc.gov/nchs/products/databriefs/db427.htm#section_5.

23 *Carbon pollution has been accumulating*: NASA, "Graphic: The Relentless Rise of Carbon Dioxide," *climate.nasa.gov*, accessed February 23, 2023, https://climate.nasa.gov/climate_resources/24/graphic-the-relentless-rise-of-carbon-dioxide/.

24 *It has forced Earth's average surface temperature*: National Oceanic and Atmospheric Administration (NOAA), "Climate Change: Global Temperature," *climate.gov*, accessed February 23, 2023, https://www.climate.gov/news-features/understanding-climate/climate-change-global-temperature.

24 *They tell us that even if current international climate pledges*: United Nations Environmental Programme, *Emissions Gap Report 2023: Broken Record - Temperatures Hit New Highs, Yet World Fails to Cut Emissions (Again)*, *unep.org*, November 20, 2023, accessed December 15, 2023, https://www.unep.org/emissions-gap-report-2023.

24 *Thanks to the falling price of renewable energy*: Chris Mooney and Andrew Freedman, "We May Avoid the Very Worst Climate Scenario. But the Next-Worst Is Still Pretty Awful," *Washington Post*, January 30, 2020, accessed February 23, 2023, https://www.washingtonpost.com/weather/2020/01/30/we-may-avoid-very-worst-climate-scenario-next-worst-is-still-pretty-awful/.

24 *Yet already, at 1.2 degrees of warming*: Richard A. Betts, Stephen E. Belcher, Leon Hermanson et al., "Approaching 1.5° C: How Will We Know We've Reached This Crucial Warming Mark?," *Nature 624* (2023): 33–35; United Nations, "Updated NDC Synthesis Report: Worrying Trends Confirmed," *unfccc.int*, October 25,

2021, accessed February 23, 2023, https://unfccc.int/news/updated-ndc-synthesis-report-worrying-trends-confirmed.

Children's Lungs Are Different

25 *As a newborn, she had only*: Matt J. Herring, Lei F. Putney, Gregory Wyatt et al., "Growth of Alveoli During Postnatal Development in Humans Based on Stereological Estimation," *American Journal of Physiology—Lung Cellular and Molecular Physiology 307*: 4 (2014): L338–L344; Matthias Ochs, Jens R. Nyengaard, Anja Jung et al., "The Number of Alveoli in the Human Lung," *American Journal of Respiratory and Critical Care Medicine 169*: 1 (2004): 120–124.

25 *By the end of adolescence*: Michael P. Combs and Robert P. Dickson, "Turning the Lungs Inside Out: The Intersecting Microbiomes of the Lungs and the Built Environment," *American Journal of Respiratory and Critical Care Medicine 202*: 12 (2020): 1618–1620.

25 *Air pollutants damage lung tissue*: Judith A. Voynow and Richard Auten, "Environmental Pollution and the Developing Lung," *Clinical Pulmonary Medicine 22*: 4 (2015): 177–184.

25 *They also appear to "turn off" genes*: Luca Ferrari, Michele Carugno, and Valentina Bollati, "Particulate Matter Exposure Shapes DNA Methylation Through the Lifespan," *Clinical Epigenetics 11*: 129 (2019): 1–14.

26 *As a result, children exposed to chronic air pollution*: W. James Gauderman, Edward Avol, Frank Gilliland et al., "The Effect of Air Pollution on Lung Development from 10 to 18 Years of Age," *New England Journal of Medicine 351*: 11 (2004): 1057–1067.

26 *If children are removed from high-pollution areas*: W. James Gauderman, Robert Urman, Edward Avol et al., "Association of Improved Air Quality with Lung Development in Children," *New England Journal of Medicine 372*: 10 (2015): 905–913; Edward L. Avol, W. James Gauderman, Sylvia M. Tan et al., "Respiratory Effects of Relocating to Areas of Differing Air Pollution Levels," *American Journal of Respiratory and Critical Care Medicine 164*: 11 (2001): 2067–2072.

26 *But finding clean air can be a challenge*: American Lung Association, *State of the Air, 2023 Report* (Chicago: American Lung Association, 2023), p. 12, accessed June 30, 2023, https://www.lung.org/getmedia/338b0c3c-6bf8-480f-9e6e-b93868c6c476/SOTA-2023.pdf.

26 *But air pollutants, including those found in wildfire smoke*: Drew A. Glencross, Tzer-Ren Ho, Nuria Camina et al., "Air Pollution and Its Effects on the Immune System," *Free Radical Biology and Medicine 151* (2020): 56–68.

26 *Children with impaired lung function*: George Russell, "Asthma and Growth," *Archives of Disease in Childhood 69*: 6 (1993): 695; James A. Blackman and Matthew J. Gurka, "Developmental and Behavioral Comorbidities of Asthma in Children," *Journal of Developmental & Behavioral Pediatrics 28*: 2 (2007): 92–99.

26 *When they become adults*: Holger J. Schunemann, Joan Dorn, Brydon J. B. Grant et al., "Pulmonary Function Is a Long-Term Predictor of Mortality in the General Population: 29-Year Follow-Up of the Buffalo Health Study," *Chest 118*: 3 (2000): 656–664; D. J. Hole, G. C. M. Watt, G. Davey-Smith et al., "Impaired Lung Function and Mortality Risk in Men and Women: Findings from the Renfrew and Paisley Prospective Population Study," *British Medical Journal 313*: 7059 (1996): 711–715.

26 *Injuries to young lungs can ripple through a lifetime*: Louis I. Landau and Lynn M.

Taussig, "Early Childhood Origins and Economic Impact of Respiratory Disease Throughout Life," in Lynn M. Taussig and Louis I. Landau, eds., *Pediatric Respiratory Medicine*, 2d ed. (Maryland Heights, MO: Mosby International, 2008), pp. 1–8.

26 *Pollutants inhaled by pregnant women*: Casper-Emil Tingskov Pedersen, Anders Ulrik Eliasen, Matthias Ketzel et al., "Prenatal Exposure to Ambient Air Pollution Is Associated with Early Life Immune Perturbations," *Journal of Allergy and Clinical Immunology 151*: 1 (2023): 212–221; Mariana Materas Veras, Nilmara de Oliveira Alves, Lais Fajersztajn et al., "Before the First Breath: Prenatal Exposures to Air Pollution and Lung Development," *Cell and Tissue Research 367*: (2017): 445–455; Sophie Michel, Aishwarya Atmakuri, and Ondine S. von Ehrenstein, "Prenatal Exposure to Ambient Air Pollutants and Congenital Heart Defects: An Umbrella Review," *Environment International 178*: (2023): 108076; Phillipe Grandjean, "Chapter 2: The Placenta Is Not a Protective Armor," *Only One Chance: How Environmental Pollution Impairs Brain Development—and How to Protect the Brains of the Next Generation* (New York: Oxford University Press, 2015), pp. 13–29.

26 *They can also stunt fetal growth*: Rakesh Ghosh, Kate Causey, Katrin Burkart et al., "Ambient and Household PM2.5 Pollution and Adverse Perinatal Outcomes: A Meta-regression and Analysis of Attributable Global Burden for 204 Countries and Territories," *PLoS Medicine 18*: 9 (2021): e1003718; Bruce Bekkar, Susan Pacheco, Rupa Basu, and Nathaniel DeNicola, "Association of Air Pollution and Heat Exposure with Preterm Birth, Low Birth Weight, and Stillbirth in the US: A Systematic Review," *JAMA Network Open 3*: 6 (2020): e208243; US Centers for Disease Control and Prevention (CDC), "Premature Birth," *cdc.gov*, last reviewed November 1, 2022, accessed February 23, 2023, https://www.cdc.gov/reproductivehealth/features/premature-birth/.

26 *Those premature babies—especially if born at 32 weeks or sooner*: Jasper V. Been, Marlies J. Lugtenberg, Eline Smets et al., "Preterm Birth and Childhood Wheezing Disorders: A Systematic Review and Meta-analysis," *PLoS Medicine 11*: 1 (2014): e1001596.

26 *Even at rest, Anna was breathing up to twice as fast*: Susannah Fleming, Matthew Thompson, Richard Stevens et al., "Normal Ranges of Heart Rate and Respiratory Rate in Children from Birth to 18 Years of Age: A Systematic Review of Observational Studies," *Lancet 377*: 9770 (2011): 1011–1018.

26 *Also, though her lung surface area was smaller*: Joel Schwartz, "Air Pollution and Children's Health," *Pediatrics 113*: Supplement_3 (2004): 1037–1043.

27 *Children play more outdoors*: Neil A. Buzzard, Nigel N. Clark, and Steven E. Guffey, "Investigation into Pedestrian Exposure to Near-Vehicle Exhaust Emissions," *Environmental Health 8*: 1 (2009): 1–13.

27 *School-age children tend to mouth breathe more*: William D. Bennett, Kirby L. Zeman, and Annie M. Jarabek, "Nasal Contribution to Breathing and Fine Particle Deposition in Children Versus Adults," *Journal of Toxicology and Environmental Health, Part A 71*: 3 (2007): 227–237.

27 *And the cells that line their airways*: Lilian Calderón-Garcidueñas, Gildardo Valencia-Salazar, Antonio Rodríguez-Alcaraz et al., "Ultrastructural Nasal Pathology in Children Chronically and Sequentially Exposed to Air Pollutants," *American Journal of Respiratory Cell and Molecular Biology 24*: 2 (2001): 132–138.

World on Fire

29 *Since the Rim Fire, more than a hundred scientific studies*: Adam J. P. Smith, Matthew W. Jones, John T. Abatzoglou et al., "Climate Change Increases the Risk of Wildfires: September 2020 Update," *sciencebrief.org*, accessed March 3, 2023, https://sciencebrief.org/uploads/reviews/ScienceBrief_Review_WILDFIRES_Sep2020.pdf.

29 *I knew why 129 million trees had died*: USFS and CalFire, "Record 129 Million Trees Dead in California."

29 *How forest managers and utility companies*: George Busenberg, "Wildfire Management in the United States: The Evolution of a Policy Failure," *Review of Policy Research 21*: 2 (2004): 145–156.

29 *Why wildfires were much bigger*: Anthony LeRoy Westerling, "Increasing Western US Forest Wildfire Activity: Sensitivity to Changes in the Timing of Spring," *Philosophical Transactions of the Royal Society B: Biological Sciences 371*: 1696 (2016): 20150178; National Interagency Fire Center, "Total Wildland Fires and Acres (1960–2015)," *nifc.gov*, accessed March 10, 2023, https://www.nifc.gov/fireInfo/fireInfo_stats_totalFires.html.

Where There's Smoke, There's Illness

30 *When I was a child, the country had never seen*: Manas Sharma, Simon Scarr, and Sharon Bernstein, "The Age of the 'Megafire,'" *Reuters*, February 1, 2021, accessed March 10, 2023, https://www.reuters.com/graphics/USA-WILDFIRES/EXTREMES/qzjvqmmravx/; Ana G. Rappold, Jeanette Reyes, George Pouliot et al., "Community Vulnerability to Health Impacts of Wildland Fire Smoke Exposure," *Environmental Science & Technology 51*: 12 (2017): 6674–6682.

31 *But during the Rim Fire, special aircraft*: Xiaoxi Liu, L. Gregory Huey, Robert J. Yokelson et al., "Airborne Measurements of Western US Wildfire Emissions: Comparison with Prescribed Burning and Air Quality Implications," *Journal of Geophysical Research: Atmospheres 122*: 11 (2017): 6108–6129.

31 *The smallest particles regulated by law*: US Environmental Protection Agency (EPA), "Why Wildfire Smoke Is a Health Concern," *epa.gov*, accessed February 23, 2023, https://www.epa.gov/wildfire-smoke-course/why-wildfire-smoke-health-concern.

31 *Bits of what once was, particles can be carried*: Amir Sapkota, J. Morel Symons, Jan Kleissl et al., "Impact of the 2002 Canadian Forest Fires on Particulate Matter Air Quality in Baltimore City," *Environmental Science & Technology 39*: 1 (2005): 24–32.

31 *Because of their minute size*: Annette Peters, Bellina Veronesi, Lilian Calderón-Garciduenas et al., "Translocation and Potential Neurological Effects of Fine and Ultrafine Particles a Critical Update," *Particle and Fibre Toxicology 3*: 1 (2006): 1–13.

31 *Fossil fuels—whether burned in*: Frederica Perrera, "Pollution from Fossil-Fuel Combustion is the Leading Environmental Threat to Global Pediatric Health and Equity: Solutions Exist," *International Journal of Environmental Research and Public Health 15*: 1 (2017): 16.

31 *Over the last half century, the Clean Air Act reduced*: EPA, "Clean Air Act Overview: Progress Cleaning the Air and Improving People's Health," *epa.gov*, last modified

May 1, 2023, accessed July 7, 2023, https://www.epa.gov/clean-air-act-overview/progress-cleaning-air-and-improving-peoples-health.

31 *Yet scientists estimate that in 2018 an astounding 13 percent*: Karn Vohra, Alina Vodonos, Joel Schwartz et al., "Global Mortality from Outdoor Fine Particle Pollution Generated by Fossil Fuel Combustion: Results from GEOS-Chem," *Environmental Research 195* (2021): 110754.

31 *Between 2017 and 2021, both the severity of smoke pollution*: American Lung Association, *State of the Air 2023*, pp.13–14.

32 *But as wildfire smoke from Canada engulfed*: Aatish Bhatia, Josh Katz, and Margot Sanger-Katz, "Just How Bad Was the Pollution in New York?" *New York Times,* June 8, 2023, updated June 9, 2023, accessed June 11, 2023, https://nytimes.com/interactive/2023/06/06/upshot/new-york-city-smoke.html; Amy Qin, Matt Kiefer, and Brett Chase, "Chicago Recorded Its Worst-Ever Fine Particle Air Pollution This Week," *wbez.org*, June 30, 2023, accessed August 26, 2023, https://www.wbez.org/stories/it-was-a-bad-week-for-air-pollution-in-chicago/653a6f8e-dc82-4e0b-ba78-4ef9cb01b933; Claire Thornton, Trevor Hughes, Jeanine Santucci, and Doyle Rice, "Washington, Chicago, Detroit Have Some of the Worst Air Quality in the World: Updates," *USA Today*, June 29, 2023, updated June 30, 2023, accessed August 26, 2023, https://www.usatoday.com/story/news/nation/2023/06/29/air-quality-canadian-wildfires-smoke-live-updates/70367926007/.

32 *As many as 3,000 Californians over the age of sixty-five*: Hugh D. Safford, Alison K. Paulson, Zachary L. Steel et al., "The 2020 California Fire Season: A Year like No Other, a Return to the Past or a Harbinger of the Future?," *Global Ecology and Biogeography 31*: 10 (2022): 2005–2025.

32 *In adults, particle pollution increases the risks of*: Jason Sacks, Barbara Buckley, Stephanie Deflorio-Barker et al., *Supplement to the 2019 Integrated Science Assessment for Particulate Matter* (Research Triangle Park, NC: EPA, 2022), *ncbi.nlm.nih.gov*, accessed March 10, 2023, https://www.ncbi.nlm.nih.gov/books/NBK588505/; Eric V. Balti, Justin B. Echouffo-Tcheugui, Yandiswa Y. Yako, and Andre P. Kengne, "Air Pollution and Risk of Type 2 Diabetes Mellitus: A Systematic Review and Meta-analysis," *Diabetes Research and Clinical Practice 106*: 2 (2014): 161–172; Wenqi Xu, Shaopeng Wang, Liping Jiang et al., "The Influence of PM2.5 Exposure on Kidney Diseases," *Human & Experimental Toxicology 41* (2022): 09603271211069982.

32 *COVID-19 appears to be more easily transmitted*: Daniel Kiser, Gai Elhanan, William J. Metcalf et al., "SARS-CoV-2 Test Positivity Rate in Reno, Nevada: Association with PM2.5 During the 2020 Wildfire Smoke Events in the Western United States," *Journal of Exposure Science & Environmental Epidemiology 31*: 5 (2021): 797–803.

32 *What is clear is that when particle levels jump*: Sacks et al., *Supplement to the 2019 Integrated Science Assessment for Particulate Matter.*

32 *But three studies that focused on the particulate matter in smoke*: Nino Kunzli, Ed Avol, Jun Wu et al., "Health Effects of the 2003 Southern California Wildfires on Children," *American Journal of Respiratory and Critical Care Medicine 174*: 11 (2006): 1221–1228; Justine A. Hutchinson, Jason Vargo, Meredith Milet et al., "The San Diego 2007 Wildfires and Medi-Cal Emergency Department Presentations, Inpatient Hospitalizations, and Outpatient Visits: An Observational Study of Smoke

Exposure Periods and a Bidirectional Case-Crossover Analysis," *PLoS Medicine 15*: 7 (2018): e1002601; Sydney Leibel, Margaret Nguyen, William Brick et al., "Increase in Pediatric Respiratory Visits Associated with Santa Ana Wind–Driven Wildfire Smoke and PM2.5 Levels in San Diego County," *Annals of the American Thoracic Society 17*: 3 (2020): 313–320.

32 *Every major wildfire leapt from their graph*: Daniel Kiser and William J. (Jim) Metcalf, personal communication, Desert Research Institute, April 18, 2019.

33 *In recent research from Stanford University*: Mary Prunicki, Rodd Kelsey, Justin Lee et al., "The Impact of Prescribed Fire Versus Wildfire on the Immune and Cardiovascular Systems of Children," *Allergy 74*: 10 (2019): 1989–1991.

33 *Particles from wildfires cause more lung inflammation*: Rosana Aguilera, Thomas Corringham, Alexander Gershunov, and Tarik Benmarhnia, "Wildfire Smoke Impacts Respiratory Health More Than Fine Particles from Other Sources: Observational Evidence from Southern California," *Nature Communications 12*: 1 (2021): 1493.

33 *In fact, a multiyear study in San Diego County*: Rosana Aguilera, Thomas Corringham, Alexander Gershunov et al., "Fine Particles in Wildfire Smoke and Pediatric Respiratory Health in California," *Pediatrics 147*: 4 (2021): e2020027128.

33 *And particles from megafires seem to inflict more damage*: Lee Ann L. Hill, Jessie M. Jaeger, and Audrey Smith, *Can Prescribed Fires Mitigate Health Harm? A Review of Air Quality and Public Health Implications of Wildfire and Prescribed Fire* (Oakland, CA: PSE Healthy Energy and American Lung Association, 2022), accessed February 23, 2023, https://www.lung.org/getmedia/fd7ff728-56d9-4b33-82eb-abd06f01bc3b/pse_wildfire-and-prescribed-fire-brief_final_2022.pdf.

33 *Take, for example, polycyclic aromatic hydrocarbons*: Frederica Perera, *Children's Health and the Peril of Climate Change* (New York: Oxford University Press, 2022), p. 33.

34 *Schools keep kids inside*: "WCSD Releases Statement About Poor Air Quality, Smoke Days," *2news.com,* August 16, 2021, last modified November 4, 2021, accessed February 23, 2023, https://www.2news.com/wcsd-releases-statement-about-poor-air-quality-smoke-days/article_a3d0f4aa-69c5-52ec-ace1-ab587fd09daa.html.

Inhaled Particles Cause Long-Term Harm

34 *I noted earlier that a child who regularly breathes*: Gauderman et al., "The Effect of Air Pollution on Lung Development from 10 to 18 Years of Age."

34 *But for young adults who had not been so lucky*: Schunemann et al., "Pulmonary Function Is a Long-Term Predictor of Mortality in the General Population: 29-Year Follow-Up of the Buffalo Health Study"; Hole et al., "Impaired Lung Function and Mortality Risk in Men and Women: Findings from the Renfrew and Paisley Prospective Population Study"; Landau and Taussig, "Early Childhood Origins and Economic Impact of Respiratory Disease Throughout Life."

35 *It's clear that low-income and minority children*: Perera, "Chapter 4: Children Are Not All Equal," in *Children's Health and the Peril of Climate Change*, pp. 80–100.

35 *Babies exposed to high particle levels*: Tracey J. Woodruff, Jennifer D. Parker, and Kenneth C. Schoendorf, "Fine Particulate Matter (PM2.5) Air Pollution and Selected Causes of Postneonatal Infant Mortality in California," *Environmental Health Per-*

spectives 114: 5 (2006): 786–790; Nihit Goyal, Mahesh Karra, and David Canning, "Early-Life Exposure to Ambient Fine Particulate Air Pollution and Infant Mortality: Pooled Evidence from 43 Low- and Middle-Income Countries," *International Journal of Epidemiology 48*: 4 (2019): 1125–1141; Ji-Young Son, Hyung Joo Lee, Petros Koutrakis, and Michelle L. Bell, "Pregnancy and Lifetime Exposure to Fine Particulate Matter and Infant Mortality in Massachusetts, 2001–2007," *American Journal of Epidemiology 186*: 11 (2017): 1268–1276.

35 *Particle pollution increases the incidence of sudden infant death syndrome*: Woodruff et al., "Fine Particulate Matter (PM2.5) Air Pollution and Selected Causes of Postneonatal Infant Mortality in California"; Sumi Mehta, Hwashin Shin, Rick Burnett et al., "Ambient Particulate Air Pollution and Acute Lower Respiratory Infections: A Systematic Review and Implications for Estimating the Global Burden of Disease," *Air Quality, Atmosphere & Health 6* (2013): 69–83; August Wrotek, Artur Badyda, Piotr O. Czechowski et al., "Air Pollutants' Concentrations Are Associated with Increased Number of RSV Hospitalizations in Polish Children," *Journal of Clinical Medicine 10*: 15 (2021): 3224.

35 *A Columbia University analysis*: Allison S. Larr and Matthew Neidell, "Pollution and Climate Change," *The Future of Children 26*: 1 (2016): 93–113.

35 *When particles are inhaled by a mother*: Keren Agay-Shay, Michael Friger, Shai Linn et al., "Air Pollution and Congenital Heart Defects," *Environmental Research 124* (2013): 28–34.

35 *Prenatal exposure has been tied to a higher risk of cancer*: Eric Lavigne, Isac Lima, Marianne Hatzopoulou et al., "Ambient Ultrafine Particle Concentrations and Incidence of Childhood Cancers," *Environment International 145* (2020): 106135; Yufan Liu, Yan Li, Hailin Xu et al., "Pre- and Postnatal Particulate Matter Exposure and Blood Pressure in Children and Adolescents: A Systematic Review and Meta-analysis," *Environmental Research 223* (2023): 115373; Simona Simkova, Milos Veleminsky, and Radim J. Sram, "The Impact of Air Pollution to Obesity," *Neuroendocrinology Letters 41*: 3 (2020): 146–153.

36 *Particles enter and inflame the placenta*: Hannelore Bové, Eva Bongaerts, Eli Slenders et al., "Ambient Black Carbon Particles Reach the Fetal Side of Human Placenta," *Nature Communications 10* (2019): 3866; Schwartz, "Air Pollution and Children's Health."

36 *While a baby can be born too early or too small*: Ghosh et al., "Ambient and Household PM2.5 Pollution and Adverse Perinatal Outcomes: A Meta-regression and Analysis of Attributable Global Burden for 204 Countries and Territories."

36 *Small, preterm babies then have*: CDC, "Reproductive Health: Preterm Birth," *cdc.gov*, last reviewed November 1, 2022, accessed February 24, 2023, https://www.cdc.gov/reproductivehealth/maternalinfanthealth/pretermbirth.htm.

36 *In fact, complications of premature birth*: World Health Organization (WHO), *Born Too Soon: Decade of Action on Preterm Birth* (Geneva: World Health Organization, 2023), p. 1.

36 *Those who live in high-particle areas*: Mengting Shang, Meng Tang, and Yuying Xue, "Neurodevelopmental Toxicity Induced by Airborne Particulate Matter," *Journal of Applied Toxicology 43*: 1 (2023): 167–185; S. Franco Suglia, Alexandros Gryparis, Robert O. Wright et al., "Association of Black Carbon with Cognition Among Children in a Prospective Birth Cohort Study," *American Journal of Epidemiology 167*: 3

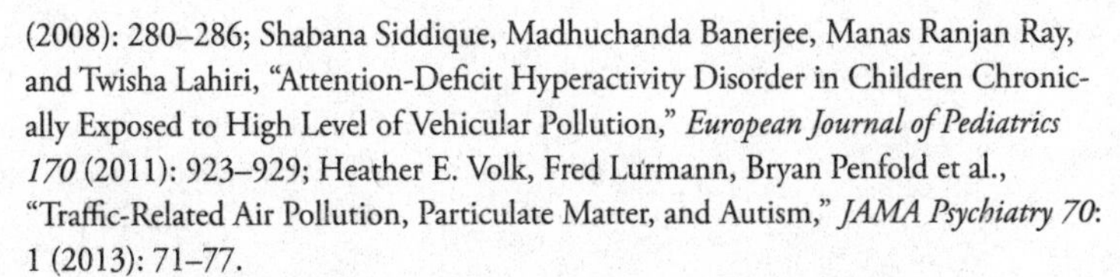
(2008): 280–286; Shabana Siddique, Madhuchanda Banerjee, Manas Ranjan Ray, and Twisha Lahiri, "Attention-Deficit Hyperactivity Disorder in Children Chronically Exposed to High Level of Vehicular Pollution," *European Journal of Pediatrics 170* (2011): 923–929; Heather E. Volk, Fred Lurmann, Bryan Penfold et al., "Traffic-Related Air Pollution, Particulate Matter, and Autism," *JAMA Psychiatry 70*: 1 (2013): 71–77.

36 *One likely but disturbing reason*: R. G. Lucchini, D. C. Dorman, A. Elder, and B. Veronesi, "Neurological Impacts from Inhalation of Pollutants and the Nose–Brain Connection," *Neurotoxicology 33*: 4 (2012): 838–841.

36 *But when scientists scanned the brains*: Lilian Calderón-Garcidueñas, Antonieta Mora-Tiscareño, Esperanza Ontiveros et al., "Air Pollution, Cognitive Deficits and Brain Abnormalities: A Pilot Study with Children and Dogs," *Brain and Cognition 68*: 2 (2008): 117–127.

36 *Later research confirmed that particles*: Lilian Calderón-Garcidueñas, Angélica Gónzalez-Maciel, Rafael Reynoso-Robles et al., "Hallmarks of Alzheimer Disease Are Evolving Relentlessly in Metropolitan Mexico City Infants, Children and Young Adults. APOE4 Carriers Have Higher Suicide Risk and Higher Odds of Reaching NFT Stage V at ≤ 40 Years of Age," *Environmental Research 164* (2018): 475–487; Barbara A. Maher, Imad A. M. Ahmed, Vassil Karloukovski et al., "Magnetite Pollution Nanoparticles in the Human Brain," *Proceedings of the National Academy of Sciences 113*: 39 (2016): 10797–10801.

36 *And it had sowed the seeds of later dementia*: Yikai Shou, Yilu Huang, Xiaozheng Zhu et al., "A Review of the Possible Associations Between Ambient PM2.5 Exposures and the Development of Alzheimer's Disease," *Ecotoxicology and Environmental Safety 174*: (2019): 344–352; Diana Younan, Andrew J. Petkus, Keith F. Widaman et al., "Particulate Matter and Episodic Memory Decline Mediated by Early Neuroanatomic Biomarkers of Alzheimer's Disease," *Brain 143*: 1 (2020): 289–302.

36 *Though these tiny pollutants*: Yu Qi, Shuting Wei, Tao Xin et al., "Passage of Exogenous Fine Particles from the Lung into the Brain in Humans and Animals," *Proceedings of the National Academy of Sciences 119*: 26 (2022): e2117083119; Lillian Calderón-Garcidueñas, Ángel Augusto Pérez-Calatayud, Angélica González-Maciel et al., "Environmental Nanoparticles Reach Human Fetal Brains," *Biomedicines 10*: 2 (2022): 410.

37 *Regardless of their path, when particles*: Lucio G. Costa, Toby B. Cole, Khoi Dao et al., "Effects of Air Pollution on the Nervous System and Its Possible Role in Neurodevelopmental and Neurodegenerative Disorders," *Pharmacology & Therapeutics 210* (2020): 107523; Bradley S. Peterson, Virginia A. Rauh, Ravi Bansal et al., "Effects of Prenatal Exposure to Air Pollutants (Polycyclic Aromatic Hydrocarbons) on the Development of Brain White Matter, Cognition, and Behavior in Later Childhood," *JAMA Psychiatry 72*: 6 (2015): 531–540.

37 *That's the theory, at least, as to why so much new research*: Perera, *Children's Health and the Peril of Climate Change*, pp. 52–56.

37 *In a multiyear study of almost 300,000 children*: Md Mostafijur Rahman, Yu-Hsiang Shu, Ting Chow et al., "Prenatal Exposure to Air Pollution and Autism Spectrum

Disorder: Sensitive Windows of Exposure and Sex Differences," *Environmental Health Perspectives 130*: 1 (2022): 017008.

37 *And a recent study of over 164,000 school-age children*: Ru-Qing Liu, Yuming Guo, Michael S. Bloom et al., "Differential Patterns of Association Between PM1 and PM2.5 with Symptoms of Attention Deficit Hyperactivity Disorder," *Nature Mental Health* (2023): 1–8.

Good Ozone/Bad Ozone

39 *Of the twenty-five most ozone-polluted cities*: American Lung Association, *State of the Air 2019* (Chicago: American Lung Association, 2019), p. 5.

39 *Between 2019 and 2021, ozone levels improved*: American Lung Association, *State of the Air 2023*, pp. 17–19.

40 *In fact, Reno's ozone levels continued to worsen*: Amy Alonzo, "American Lung Association: Reno's Air Quality Among Worst in Nation," *Reno Gazette Journal*, April 19, 2023, accessed September 6, 2023, https://www.rgj.com/story/news/2023/04/19/american-lung-association-renos-air-quality-among-worst-in-the-nation/70126847007/.

40 *The Rim Fire generated air pollution equivalent to*: Sierra Nevada Conservancy, "The Rim Fire: Why Investing in Forest Health Equals Investing in the Health of California," *sierranevada.ca.gov*, October 31, 2013, accessed February 25, 2023, https://www.hcd.ca.gov/community-development/disaster-recovery-programs/ndrc-application-documents/docs/10.31rimfirefactsheet.pdf.

40 *When those gases mixed in the fire's heat*: Dan Jaffe, Duli Chand, Will Hafner et al., "Influence of Fires on O3 Concentrations in the Western US," *Environmental Science & Technology 42*: 16 (2008): 5885–5891.

40 *It's estimated that over two thousand pediatric emergency room visits*: Jacob R. Pratt, Ryan W. Gan, Bonne Ford et al., "A National Burden Assessment of Estimated Pediatric Asthma Emergency Department Visits That May Be Attributed to Elevated Ozone Levels Associated with the Presence of Smoke," *Environmental Monitoring and Assessment 191*: 269 (2019): 1–13.

40 *For her, even a small increase in ozone*: Wanting Huang, Jinzhun Wu, and Xiaoliang Lin, "Ozone Exposure and Asthma Attack in Children," *Frontiers in Pediatrics 10* (2022): 830897.

40 *The evidence for this comes from studies*: Rob McConnell, Kiros Berhane, Frank Gilliland et al., "Asthma in Exercising Children Exposed to Ozone: A Cohort Study," *Lancet 359*: 9304 (2002): 386–391; Rob McConnell, Kiros Berhane, Ling Yao et al., "Traffic, Susceptibility, and Childhood Asthma," *Environmental Health Perspectives 114*: 5 (2006): 766–772.

40 *Like the particulate matter Liam inhaled*: Stephanie M. Holm and John R. Balmes, "Systematic Review of Ozone Effects on Human Lung Function, 2013 Through 2020," *Chest 161*: 1 (2022): 190–201.

41 *It may also affect his learning*: Frank D. Gilliland, Kiros Berhane, Edward B. Rappaport et al., "The Effects of Ambient Air Pollution on School Absenteeism Due to Respiratory Illnesses," *Epidemiology 12*: 1 (2001): 43–54.

41 *As was the case with particle pollution*: Beate Ritz, Fei Yu, Scott Fruin et al., "Ambient

Air Pollution and Risk of Birth Defects in Southern California," *American Journal of Epidemiology 155*: 1 (2002): 17–25; Bing-Fang Hwang and Jouni J. K. Jaakkola, "Ozone and Other Air Pollutants and the Risk of Oral Clefts," *Environmental Health Perspectives 116*: 10 (2008): 1411–1415; Muhammad T. Salam, Joshua Millstein, Yu-Fen Li et al., "Birth Outcomes and Prenatal Exposure to Ozone, Carbon Monoxide, and Particulate Matter: Results from the Children's Health Study," *Environmental Health Perspectives 113*: 11 (2005): 1638–1644; Silvia Regina Dias Médici Saldiva, Ligia Vizeu Barrozo, Cléa Rodrigues Leone et al., "Small-Scale Variations in Urban Air Pollution Levels Are Significantly Associated with Premature Births: A Case Study in São Paulo, Brazil," *International Journal of Environmental Research and Public Health 15*: 10 (2018): 2236; Nazeeba Siddika, Aino K. Rantala, Harri Antikainen et al., "Synergistic Effects of Prenatal Exposure to Fine Particulate Matter (PM2.5) and Ozone (O3) on the Risk of Preterm Birth: A Population-Based Cohort Study," *Environmental Research 176* (2019): 108549; Tracey J. Woodruff, Lyndsey A. Darrow, and Jennifer D. Parker, "Air Pollution and Postneonatal Infant Mortality in the United States, 1999–2002," *Environmental Health Perspectives 116*: 1 (2008): 110–115; Junfeng Zhang, Yongjie Wei, and Zhangfu Fang, "Ozone Pollution: A Major Health Hazard Worldwide," *Frontiers in Immunology 10* (2019): 2518.

Medical Mysteries

42 *Though fatal attacks are rare*: CDC, "Asthma Data, Statistics, and Surveillance, 2020 Archived National Asthma Data: Mortality," *cdc.gov*, accessed February 25, 2023, https://www.cdc.gov/asthma/archivedata/2020/2020_archived_national_data.html.

42 *The statistics had been pounded into me*: Lara J. Akinbami, Jeanne E. Moorman, Paul L. Garbe, and Edward J. Sondik, "Status of Childhood Asthma in the United States, 1980–2007," *Pediatrics 123*, Supplement_3 (2009): S131–S145.

42 *Kids were more likely than adults to get the disease*: Cynthia A. Pate, Hatice S. Zahran, Xiaoting Qin et al., "Asthma Surveillance—United States, 2006–2018," *Morbidity and Mortality Weekly Report, Surveillance Summaries 70*: 5 (2021): 1; Rachel A. Winer, Xiaoting Qin, Theresa Harrington et al., "Asthma Incidence Among Children and Adults: Findings from the Behavioral Risk Factor Surveillance System Asthma Call-Back Survey—United States, 2006–2008," *Journal of Asthma 49*: 1 (2012): 16–22.

42 *Until the 1960s, some doctors had believed*: Kenneth Purcell, "Distinctions Between Subgroups of Asthmatic Children: Children's Perceptions of Events Associated with Asthma," *Pediatrics 31*: 3 (1963): 486–494 ("suppressed cry" is on p. 492); Thomas Morton French and Franz Alexander, "Psychogenic Factors in Bronchial Asthma," *Psychosomatic Medicine Monographs 1*: 4 (Washington, DC: National Research Council, 1941).

42 *Not until the 1980s did we begin to understand*: E. R. McFadden Jr., "A Century of Asthma," *American Journal of Respiratory and Critical Care Medicine 170*: 3 (2004): 215–221; Stephen T. Holgate, "A Brief History of Asthma and Its Mechanisms to Modern Concepts of Disease Pathogenesis," *Allergy, Asthma & Immunology Research 2*: 3 (2010): 165–171.

43 *Handheld inhalers to deliver asthma medication*: Stephen W. Stein and Charles G. Thiel, "The History of Therapeutic Aerosols: A Chronological Review," *Journal of Aerosol Medicine and Pulmonary Drug Delivery 30*: 1 (2017): 20–41.

43 *"Prior to 1960," one medical historian noted*: Thomas A. E. Platts-Mills, "The Allergy Epidemics: 1870–2010," *Journal of Allergy and Clinical Immunology 136*: 1 (2015): 3–13.

43 *The percentage of American children with the disease*: Jeanne E. Moorman, Lara J. Akinbami, Cathy M. Bailey et al., "National Surveillance of Asthma: United States, 2001–2010," *Vital & Health Statistics, Series 3, Analytical and Epidemiological Studies 35* (2012): 1–58.

43 *All* atopic diseases: Kristen D. Jackson, LaJeana D. Howie, Lara J. Akinbami, "Trends in Allergic Conditions Among Children: United States, 1997–2011, NCHS Data Brief No. 121," National Center for Health Statistics (2013), *cdc.gov/nchs*, accessed March 3, 2023, https://www.cdc.gov/nchs/data/databriefs/db121.pdf; Akinbami et al., "Status of Childhood Asthma in the United States, 1980–2007"; Christine C. Johnson, Suzanne L. Havstad, Dennis R. Ownby et al., "Pediatric Asthma Incidence Rates in the United States from 1980 to 2017," *Journal of Allergy and Clinical Immunology 148*: 5 (2021): 1270–1280.

43 *Most asthmatics have some type of allergy*: E. J. O'Connell, "The Burden of Atopy and Asthma in Children," *Allergy 59*: s78 (2004): 7–11; Michael Schatz and Lanny Rosenwasser, "The Allergic Asthma Phenotype," *Journal of Allergy and Clinical Immunology: In Practice 2*: 6 (2014): 645–648.

43 *Allergies are, like asthma, very common in children*: Benjamin Zablotsky, Lindsey I. Black, and Lara J. Akinbami, "Diagnosed Allergic Conditions in Children Aged 0–17 Years: United States, 2021, NCHS Data Brief No. 459," National Center for Health Statistics (2023), *cdc.gov/nchs*, accessed March 3, 2023, https://www.cdc.gov/nchs/products/databriefs/db459.htm.

43 *The World Allergy Organization estimates*: Ruby Pawankar, Giorgio W. Canonica, Stephen T. Holgate et al., eds., *WAO White Book on Allergy: Update 2013* (Milwaukee, WI: World Allergy Organization, 2013), p. 18; Päivi M. Salo, Samuel J. Arbes Jr., Renee Jaramillo et al., "Prevalence of Allergic Sensitization in the United States: Results from the National Health and Nutrition Examination Survey (NHANES) 2005–2006," *Journal of Allergy and Clinical Immunology 134*: 2 (2014), 350–359.

44 *Atopic diseases are less common in developing countries and rural areas*: Pawankar et al., *WAO White Book on Allergy: Update 2013*, p. 16; Alejandro Rodriguez, Elizabeth Brickley, Laura Rodrigues et al., "Urbanisation and Asthma in Low-Income and Middle-Income Countries: A Systematic Review of the Urban–Rural Differences in Asthma Prevalence," *Thorax 74*: 11 (2019): 1020–1030; Philip J. Cooper, Martha E. Chico, Maritza G. Vaca et al., "Risk Factors for Asthma and Allergy Associated with Urban Migration: Background and Methodology of a Cross-Sectional Study in Afro-Ecuadorian School Children in Northeastern Ecuador (Esmeraldas-SCAALA Study)," *BMC Pulmonary Medicine 6* (2006): 1–9.

44 *Before clean water and public sanitation*: Platts-Mills, "The Allergy Epidemics: 1870–2010."

44 *This* hygiene hypothesis *argues*: David P. Strachan, "Hay Fever, Hygiene, and Household Size," *British Medical Journal 299*: 6710 (1989): 1259–1260; Marsha Wills-Karp, Joanna Santeliz, and Christopher L. Karp, "The Germless Theory of Allergic Disease: Revisiting the Hygiene Hypothesis," *Nature Reviews Immunology 1*: 1 (2001): 69–75.

44 *Those fuels, running our cities*: Frederica Perera, "Multiple Threats to Child Health from Fossil Fuel Combustion: Impacts of Air Pollution and Climate Change," *Environmental Health Perspectives 125*: 2 (2017): 141–148.

44 *Viewed through this lens, climbing asthma rates*: Iowa State University, Iowa Community Indicators Program, "Urban Percentage of the Population for States, Historical," *icip.iastate.edu*, accessed February 26, 2023, https://www.icip.iastate.edu/tables/population/urban-pct-states.

45 *Among children like Ruby*: Antonella Zanobetti, Patrick H. Ryan, Brent Coull et al., "Childhood Asthma Incidence, Early and Persistent Wheeze, and Neighborhood Socioeconomic Factors in the ECHO/CREW Consortium," *JAMA Pediatrics 176*: 8 (2022): 759–767; Lara J. Akinbami, Jeanne E. Moorman, Cathy Bailey et al., "Trends in Asthma Prevalence, Health Care Use, and Mortality in the United States, 2001–2010, NCHS Data Brief No. 94," National Center for Health Statistics (2012), *cdc.gov/nchs*, accessed February 26, 2023, https://www.cdc.gov/nchs/data/databriefs/db94.pdf.

45 *Because of the poor-quality air*: Giulia Cesaroni, Sara Farchi, Marina Davoli et al., "Individual and Area-Based Indicators of Socioeconomic Status and Childhood Asthma," *European Respiratory Journal 22*: 4 (2003): 619–624; Cassandra R. O'Lenick, Andrea Winquist, James A. Mulholland et al., "Assessment of Neighbourhood-Level Socioeconomic Status as a Modifier of Air Pollution–Asthma Associations Among Children in Atlanta," *Journal of Epidemiology and Community Health 71*: 2 (2017): 129–136; Allison J. Burbank and David B. Peden, "Assessing the Impact of Air Pollution on Childhood Asthma Morbidity: How, When and What to Do," *Current Opinion in Allergy and Clinical Immunology 18*: 2 (2018): 124–131.

45 *Cities and suburbs create*: Mark Z. Jacobson, "Enhancement of Local Air Pollution by Urban CO_2 Domes," *Environmental Science & Technology 44*: 7 (2010): 2497–2502.

45 *And they are not only warmer*: Zihan Liu, Wenfeng Zhan, Benjamin Bechtel et al., "Surface Warming in Global Cities Is Substantially More Rapid Than in Rural Background Areas," *Communications Earth & Environment 3*: 1 (2022): 219.

45 *Consider ragweed*: Lewis H. Ziska and Frances A. Caulfield, "Rising CO2 and Pollen Production of Common Ragweed (*Ambrosia artemisiifolia L.*), a Known Allergy-Inducing Species: Implications for Public Health," *Functional Plant Biology 27*: 10 (2000): 893–898; Ben D. Singer, Lewis H. Ziska, David A. Frenz et al., "Increasing Amb a 1 Content in Common Ragweed (*Ambrosia artemisiifolia*) Pollen as a Function of Rising Atmospheric CO_2 Concentration," *Functional Plant Biology 32*: 7 (2005): 667–670; Lewis H. Ziska, Dennis E. Gebhard, David A. Frenz et al., "Cities as Harbingers of Climate Change: Common Ragweed, Urbanization, and Public Health," *Journal of Allergy and Clinical Immunology 111*: 2 (2003): 290–295; Lewis Ziska, Kim Knowlton, Christine Rogers et al., "Recent Warming by Latitude Associated with Increased Length of Ragweed Pollen Season in Central North America," *Proceedings of the National Academy of Sciences 108*: 10 (2011): 4248–4251.

45 *The "poison" in poison ivy*: Jacqueline E. Mohan, Lewis H. Ziska, William H. Schlesinger et al., "Biomass and Toxicity Responses of Poison Ivy (*Toxicodendron radicans*) to Elevated Atmospheric CO_2," *Proceedings of the National Academy of Sciences 103*: 24 (2006): 9086–9089.

45 *Allergic reactions to insect stings*: Jeffrey G. Demain, Bradford D. Gessner, Joseph B. McLaughlin et al., "Increasing Insect Reactions in Alaska: Is This Related to Changing Climate?," *Allergy and Asthma Proceedings 30*: 3 (2009): 238–243.

45 *Tree species such as oak and hickory*: USFS, "Climate Change Tree Atlas," *fs.usda.gov*, accessed February 26, 2023, https://www.fs.usda.gov/ccrc/tool/climate-change-tree-atlas.

45 *Mold is proliferating*: Erik Vance, "Heavy Rains and Hurricanes Clear a Path for Supercharged Mold," *scientificamerican.com*, December 4, 2018, accessed February 26, 2023, https://www.scientificamerican.com/article/heavy-rains-and-hurricanes-clear-a-path-for-supercharged-mold/.

46 *Even the rise in food allergies*: Amy M. Branum and Susan L. Lukacs, "Food Allergy Among US Children: Trends in Prevalence and Hospitalizations, NCHS Data Brief No. 10," National Center for Health Statistics (2008), *cdc.gov/nchs*, accessed February 26, 2023, https://www.cdc.gov/nchs/products/databriefs/db10.htm.

46 *Some scientists argue that the spike in peanut allergy*: Paul John Beggs and Nicole Ewa Walczyk, "Impacts of Climate Change on Plant Food Allergens: A Previously Unrecognized Threat to Human Health," *Air Quality, Atmosphere & Health 1* (2008): 119–123.

46 *Studies indicate that up to 43 percent of children*: You Hoon Jeon, "Pollen-Food Allergy Syndrome in Children," *Clinical and Experimental Pediatrics 63*: 12 (2020): 463–468.

46 *Pollen bound to particulates*: Farnaz Sedghy, Abdol-Reza Varasteh, Mojtaba Sankian, and Malihe Moghadam, "Interaction Between Air Pollutants and Pollen Grains: The Role on the Rising Trend in Allergy," *Reports of Biochemistry & Molecular Biology 6*: 2 (2018): 219–224.

46 *Allergens may cause a more intense response*: R. Jörres, Dennis Nowak, and Helgo Magnussen, "The Effect of Ozone Exposure on Allergen Responsiveness in Subjects with Asthma or Rhinitis," *American Journal of Respiratory and Critical Care Medicine 153*: 1 (1996): 56–64; Barbara Vagaggini, Mauro Taccola, Silvana Cianchetti et al., "Ozone Exposure Increases Eosinophilic Airway Response Induced by Previous Allergen Challenge," *American Journal of Respiratory and Critical Care Medicine 166*: 8 (2002): 1073–1077.

46 *These synergistic reactions are another reason*: Paul C. Schröder, Jing Li, Gary W. K. Wong, and Bianca Schaub, "The Rural–Urban Enigma of Allergy: What Can We Learn from Studies Around the World?," *Pediatric Allergy and Immunology 26*: 2 (2015): 95–102.

Anna's Future

46 *In May 1847 a young survivor of the Donner Party*: Mullen Jr. et al., *The Donner Party Chronicles: A Day-by-Day Account of a Doomed Wagon Train, 1846–1847*, p. 352.

Chapter Two: Silent and Invisible

49 *The department had almost 280 of these calls*: Captain Reda Riddle-Biggler, personal communication, Phoenix Fire Department, Phoenix, AZ, June 5, 2017.

50 *"This is our natural disaster"*: Brian Flom, personal communication, May 21, 2017.

Extremes

52 *I had flown from Reno, the fastest-warming city*: Climate Central, "Earth Day: US Warming Rankings."

52 *In their own childhoods, whole summers had passed*: NOAA, NWS, "Climate: NOWData—NOAA Online Weather Data: Reno Area: Daily Data for a Month: July 2018," and "Daily Data for a Month: June, July, August 1990," *weather.gov*, accessed March 3, 2023, https://www.weather.gov/wrh/climate?wfo=rev.

52 *No American city has more "extreme heat" days*: Linda Lam, "The Triple-Digit Club: Here's How Often Your City Reached 100 Degrees," *wunderground.com*, June 7, 2017, accessed July 3, 2023, https://www.wunderground.com/article/news/climate/news/100-degree-temperatures-us-cities-average-most-records; NOAA, NWS, "Climate: NOWData—NOAA Online Weather Data: Phoenix Area," *weather.gov*, accessed July 3, 2023, https://www.weather.gov/wrh/climate?wfo=psr.

52 *Now imagine that bell curve being pushed*: Nadja Popovich and Adam Pearce, "It's Not Your Imagination, Summers Are Getting Hotter," *New York Times*, July 28, 2017, accessed March 10, 2023, https://www.nytimes.com/interactive/2021/climate/extreme-summer-heat.html.

52 *The five-year period leading up to Cody's hike*: Weldon B. Johnson, "2016 Was Third Hottest Year for Phoenix," *Arizona Republic*, January 6, 2017, accessed March 10, 2023, https://www.azcentral.com/story/news/local/phoenix-weather/2017/01/06/2016-third-hottest-year-phoenix/96132860/.

52 *It was capped, three days before*: Climate Central, "Hottest Trending Cities," *climate central.org*, July 18, 2016, accessed January 31, 2021, https://www.climatecentral.org/climate-matters/hottest-trending-cities.

52 *Officially, about 67,500 people*: CDC, "Heat & Health Tracker," *ephtracking.cdc.gov*, accessed March 3, 2023, https://ephtracking.cdc.gov/Applications/heatTracker/; NOAA, "Weather Related Fatality and Injury Statistics," *weather.gov*, accessed March 3, 2023, https://www.weather.gov/hazstat/.

53 *Arizonans die from heat at up to 7 times*: CDC, "Heat-Related Mortality—Arizona, 1993–2002, and United States, 1979–2002," *Morbidity and Mortality Weekly Report 54*: 25 (2005): 628–630.

53 *In a single weekend in June*: Alexis Egeland, "Missing German Hiker Found Dead in Tucson; Weekend Heat Toll Rises to 6," *Arizona Republic*, June 21, 2016, accessed July 8, 2023, https://www.azcentral.com/story/news/local/arizona-breaking/2016/06/21/missing-german-hiker-found-dead-tucson-weekend-heat-toll-rises-6/86215376/.

53 *In Maricopa County alone*: Maricopa County Public Health, "Heat-Associated Deaths in Maricopa County, AZ—Final Report for 2016," accessed March 3, 2023, https://www.maricopa.gov/ArchiveCenter/ViewFile/Item/3084.

53 *Perhaps not coincidentally*: NASA, "NASA, NOAA Data Show 2016 Warmest Year on Record Globally," *nasa.gov*, January 18, 2017, accessed March 3, 2023, https://www.nasa.gov/press-release/nasa-noaa-data-show-2016-warmest-year-on-record-globally.

53 *In a place where summers can already reach*: NOAA, NWS, "Historical Extreme Temperatures in Phoenix and Yuma," *weather.gov*, accessed March 3, 2023, https://www.weather.gov/psr/ExtremeTemps.

53 *The city "could regularly hit the 130s"*: Jerry Adler, "The Reality of a Hotter World Is Already Here," *smithsonianmagazine.com*, May 2014, accessed July 3, 2023, https://www.smithsonianmag.com/science-nature/reality-hotter-world-already-here-180951172/.

53 *In June 2021, the world was stunned*: Rachel H. White, Sam Anderson, James F. Booth et al., "The Unprecedented Pacific Northwest Heatwave of June 2021," *Nature Communications 14*: 1 (2023): 727.

53 *Though it may seem like Earth's future warming is inevitable*: Hertsgaard et al., "How a Little-Discussed Revision of Climate Science Could Help Avert Doom."

53 *By 2100, when that child is elderly*: Camilo Mora, Bénédicte Dousset, Iain R. Caldwell et al., "Global Risk of Deadly Heat," *Nature Climate Change 7*: 7 (2017): 501–506.

53 *The world our children and grandchildren inherit*: Chelsea Harvey, "A Third of the World's People Already Face Deadly Heat Waves. It Could Be Nearly Three-Quarters by 2100," *Washington Post*, June 19, 2017, accessed March 3, 2023, https://www.washingtonpost.com/news/energy-environment/wp/2017/06/19/a-third-of-the-world-already-faces-deadly-heat-waves-it-could-be-nearly-three-quarters-by-2100/#.

55 *On July 19, the day after climate scientists*: Climate Central (July 18, 2016), "Hottest Trending Cities."

55 *Two days later, the National Weather Service*: Graig Graziosi, "Heat, Ozone Warnings in Effect for Valley," *Arizona Republic*, July 21, 2016, accessed March 3, 2023, https://www.azcentral.com/story/news/local/phoenix-weather/2016/07/22/phoenix-heat-ozone-warnings-issued-into-weekend/87424190/.

55 *"It was a spring without voices"*: Rachel Carson, *Silent Spring* (New York: Fawcett Crest, 1962), p. 14.

56 *"A grim specter has crept upon us"*: Ibid., pp. 14–15.

56 *The Phoenix area is home to 4.8 million people*: US Census Bureau, "Metropolitan and Micropolitan Statistical Areas Population Totals and Components of Change: 2010–2019," *census.gov*, last reviewed October 8, 2021, accessed March 3, 2023, https://www.census.gov/content/census/en/data/datasets/time-series/demo/popest/2010s-total-metro-and-micro-statistical-areas.html.

56 *Without fast action, America's fifth-largest city*: Jack Healy, "No Large City Grew Faster Than Phoenix," *New York Times*, August 12, 2021, accessed March 3, 2023, https://www.nytimes.com/2021/08/12/us/phoenix-census-fastest-growing-city.html.

"Nature's Sanatorium"

56 *In 1919, a young California couple*: Sunnyslope Historical Society & Museum, "Sunnyslope's Angels of the Desert," *sunnyslopehistoricalsociety.org*, accessed March 3, 2023, https://sunnyslopehistoricalsociety.org/brief-history-of-sunnyslope/sunnyslopes-angels-of-the-desert/.

56 *The Colleys were "healthseekers"*: Robert E. Kravetz and Alex Jay Kimmelman, *Healthseekers in Arizona* (Phoenix, AZ: Academy of Medical Sciences, Maricopa Medical Society, 1998).

57 *In the nineteenth century the disease killed more people*: Harvard University Library Open Collections Program, "Tuberculosis in Europe and North America, 1800–1922," *curiosity.lib.harvard.edu*, accessed September 27, 2017, https://curiosity.lib

.harvard.edu/contagion/feature/tuberculosis-in-europe-and-north-america-1800-1922.

57 *In 1918–1919, the Spanish influenza pandemic*: In the first two years of the COVID-19 pandemic (2020–2021), 849,000 Americans deaths (out of 331.5 million people) were attributed to the virus, representing 0.2 percent of the population; the 1918–1919 influenza pandemic struck a US population of 103 million, killing 0.6 percent of the population. CDC, "Deaths Attributed to COVID-19 on Death Certificates," accessed July 7, 2023, https://www.cdc.gov/nchs/covid19/mortality-overview.htm; Nancy Bristow, *American Pandemic: The Lost Worlds of the 1918 Influenza Epidemic* (Oxford, UK: Oxford University Press, 2012), pp. 3–4.

57 *Hippocrates, 2,500 years before*: Jacques Jouanna, "Hippocrates and the Birth of the Human Sciences," in *Hippocrates*, translated by Malcolm B. DeBevoise (Baltimore: Johns Hopkins University Press, 1999), pp. 210–242.

57 *His theories resurfaced in the eighteenth century*: Andrea Rusnock, "Hippocrates, Bacon, and Medical Meteorology at the Royal Society, 1700–1750," in David Cantor, ed., *Reinventing Hippocrates* (Milton Park, UK: Routledge, 2017), pp. 136–154.

58 *In 1885 it led to the establishment*: Stephen H. Gehlbach, "Adirondack Cure: Consumption and Edward Trudeau," in *American Plagues: Lessons from Our Battles with Disease* (Lanham, MD: Rowman & Littlefield, 2016), pp. 101–128.

58 *Sanatoriums spread rapidly*: Sheila M. Rothman, *Living in the Shadow of Death: Tuberculosis and the Social Experience of Illness in American History* (New York: Basic Books, 1994), pp. 8–22.

58 *"Nature creates and maintains"*: F. H. Redewill, "Desert Inn," *University of Arizona Exhibits: Sanatoria Pamphlets* (ca. 1910), accessed March 3, 2023, http://ualibr-exhibits.s3-website-us-west-2.amazonaws.com/pams/health.html.

58 *As testimonials spread*: Kravetz and Kimmelman, *Healthseekers in Arizona*, p. iv.

58 *Marguerite Colley and her family settled*: Ibid., pp. 51–53.

59 *By the 1920s she was running a small medical clinic*: Sunnyslope Historical Society & Museum, "Sunnyslope's Angels of the Desert."

59 *In one photograph, she looks serious*: Exhibit at North Mountain Visitor Center Museum, Phoenix, Arizona, May 21, 2017.

59 *There is another photograph of Marguerite*: Exhibit and materials at Sunnyslope Historical Society & Museum, May 20, 2017.

59 *In the 1930s, a wealthy migrant to Sunnyslope*: Ibid.

Joey

62 Teen spent year recovering after near-death hike: Mary Ellen Resendez, "Valley Boy Spent Nearly a Year Recovering After Suffering Heat-Related Illness While Hiking," *ABC15 Arizona*, July 25, 2016, accessed June 1, 2017, https://www.abc15.com/news/region-phoenix-metro/central-phoenix/valley-boy-spent-nearly-a-year-recovering-after-suffering-heat-related-illness-while-hiking; Jared Dillingham, "Teenage Heat Stroke Survivor Warns Hikers," *3TV/CBS5 Phoenix*, July 25, 2016, accessed June 1, 2017.

The Body's Furnace

63 *The answer lay in Joey's own* metabolism: The research supporting "The Body's Furnace" section is summarized by: C. Bruce Wenger, "The Regulation of Body Temperature," in *Medical Physiology* (New York: Little, Brown, and Co., 1995), pp. 527–550, and Matthew N. Cramer, Daniel Gagnon, Orlando Laitano, and Craig G. Crandall, "Human Temperature Regulation Under Heat Stress in Health, Disease, and Injury," *Physiological Reviews 102*: 4 (2022): 1907–1989.

The Fiercely Protected Range

66 *"Humans live their entire lives"*: W. Larry Kenney, "Heat Flux and Storage in Hot Environments," *International Journal of Sports Medicine 19*: Supplement_2 (1998): S92–S95.

67 *When people are exposed*: Wenger, "The Regulation of Body Temperature," pp. 542–543.

67 *It is the reason that tourists in Arizona*: Sally Ann Iverson, Aaron Gettel, Carla P. Bezold et al., "Heat-Associated Mortality in a Hot Climate: Maricopa County, Arizona, 2006–2016," *Public Health Reports 135*: 5 (2020): 631–639.

67 *Many public health researchers*: Elizabeth G. Hanna and Peter W. Tait, "Limitations to Thermoregulation and Acclimatization Challenge Human Adaptation to Global Warming," *International Journal of Environmental Research and Public Health 12*: 7 (2015): 8034–8074.

67 *But at some point, between the atmosphere of Earth*: NASA, "Solar System Exploration—Our Galactic Neighborhood: Venus," *solarsystem.nasa.gov*, accessed June 30, 2023, https://solarsystem.nasa.gov/planets/venus/overview/.

Children Are Different (Again)

68 *Like a growing number of medical schools*: Zeina Mohammed, "Harvard Medical School Votes to Embed Climate Change in Its Curriculum," *Boston Globe*, January 20, 2023, accessed March 10, 2023, https://www.bostonglobe.com/2023/01/20/metro/harvard-medical-school-votes-embed-climate-change-its-curriculum/.

68 *Much of our understanding of how heat affects us*: Jennifer Vanos, personal communication, June 15, 2017.

68 *Pound for pound, a newborn has*: Janusz Feber and Hana Krásničanová, "Measures of Body Surface Area in Children," in *Handbook of Anthropometry: Physical Measures of Human Form in Health and Disease* (New York: Springer, 2012), pp. 1249–1256.

68 *Children also produce more metabolic heat*: Wenger, "The Regulation of Body Temperature," pp. 531–532.

69 *Toddlers' sweat glands are smaller*: Bareket Falk, "Effects of Thermal Stress During Rest and Exercise in the Paediatric Population," *Sports Medicine 25* (1998): 221–240.

Protecting Children—and Childhood

70 *She found that Phoenix playgrounds often*: Jennifer K. Vanos, Ariane Middel, Grant R. McKercher et al., "Hot Playgrounds and Children's Health: A Multiscale Analysis of

Surface Temperatures in Arizona, USA," *Landscape and Urban Planning 146* (2016): 29–42.

70 *Studies have shown that the more time children spend in nature*: Amber L. Fyfe-Johnson, Marnie F. Hazlehurst, Sara P. Perrins et al., "Nature and Children's Health: A Systematic Review," *Pediatrics 148*: 4 (2021): 72–94.

70 *Cities can be 15 to 20 degrees warmer*: National Integrated Heat Health Information System, "About Urban Heat Islands," *heat.gov*, accessed July 2, 2023, https://www.heat.gov/pages/urban-heat-islands.

70 *Low-income neighborhoods are significantly hotter*: Yi Yin, Liyin He, Paul O. Wennberg, and Christian Frankenberg, "Unequal Exposure to Heatwaves in Los Angeles: Impact of Uneven Green Spaces," *Science Advances 9*: 17 (2023): eade8501.

Heat's Hidden Toll

71 *Unfortunately, increasing humidity and urbanization*: Aatish Bhata and Winston Choi-Schagrin, "Why Record-Breaking Overnight Temperatures Are So Concerning," *New York Times*, July 9, 2021, accessed July 2, 2023, https://www.nytimes.com/2021/07/09/upshot/record-breaking-hot-weather-at-night-deaths.html; NOAA, NWS, "Phoenix, AZ, Climate and Past Weather, NOWData—NOAA Online Weather Data: Monthly Summarized Data: Year Range 1970–2023: Min Temperature and Max Temperature," *weather.gov*, accessed March 3, 2023, https://www.weather.gov/wrh/Climate?wfo=psr.

71 *In fact, the city experienced its highest-ever*: April Warnecke, "Phoenix Breaks Record for Highest Overnight Low Temperature of 97 Degrees on Tuesday," *azfamily.com*, July 19, 2023, accessed September 3, 2023, https://www.azfamily.com/2023/07/19/record-heat-again-today-phoenix/.

71 *For this reason, public health officials assume*: Vjollca Berisha, senior epidemiologist for Maricopa County Public Health, personal communication, January 19, 2018.

71 *From 2006 to 2010, physicians ascribed*: Jeffrey Berko, Deborah Ingram, Shubhaya Saha, and Jennifer D. Parker, "Deaths Attributed to Heat, Cold, and Other Weather Events in the United States, 2006–2010, National Health Statistics Reports No. 76," National Center for Health Statistics (2014), *cdc.gov/nchs*, accessed September 3, 2023, https://www.cdc.gov/nchs/data/nhsr/nhsr076.pdf; Ekta Choudhary and Ambarish Vaidyanathan, "Heat Stress Illness Hospitalizations—Environmental Public Health Tracking Program, 20 States, 2001–2010," *Morbidity and Mortality Weekly Report, Surveillance Summaries 63*: 13 (2014): 1–10.

72 *Using these methods, we know*: Xavier Basagaña, Claudio Sartini, Jose Barrera-Gómez et al., "Heat Waves and Cause-Specific Mortality at All Ages," *Epidemiology 22*: 6 (2011): 765–772; G. Brooke Anderson and Michelle L. Bell, "Heat Waves in the United States: Mortality Risk During Heat Waves and Effect Modification by Heat Wave Characteristics in 43 US Communities," *Environmental Health Perspectives 119*: 2 (2011): 210–218.

72 *Several large analyses of this type*: Zhiwei Xu, Ruth A. Etzel, Hong Su et al., "Impact of Ambient Temperature on Children's Health: A Systematic Review," *Environmental Research 117* (2012): 120–131; Michael A. McGeehin and Maria Mirabelli, "The Potential Impacts of Climate Variability and Change on Temperature-Related

Morbidity and Mortality in the United States," *Environmental Health Perspectives 109*: Supplement_2 (2001): 185–189; Sarah Chapman, Cathryn E. Birch, John H. Marsham et al., "Past and Projected Climate Change Impacts on Heat-Related Child Mortality in Africa," *Environmental Research Letters 17*: 7 (2022): 074028.

72 *If the little girl at the bus stop had a grandfather*: Stefanie Kolb, Katja Radon, Marie-France Valois et al., "The Short-Term Influence of Weather on Daily Mortality in Congestive Heart Failure," *Archives of Environmental & Occupational Health 62*: 4 (2007): 169–176.

72 *Her baby brother would be more*: Nathalie Auger, William D. Fraser, Audrey Smargiassi, and Tom Kosatsky, "Ambient Heat and Sudden Infant Death: A Case-Crossover Study Spanning 30 Years in Montreal, Canada," *Environmental Health Perspectives 123*: 7 (2015): 712–716.

72 *If her mother were pregnant*: Yuval Baharav, Lilly Nichols, Anya Wahal et al., "The Impact of Extreme Heat Exposure on Pregnant People and Neonates: A State of the Science Review," *Journal of Midwifery & Women's Health 68*: 3 (2023): 324–332.

72 *She would also be more likely to lose the pregnancy*: Ibid.

72 *If the heat wave occurred during the first trimester*: John M. Graham Jr., "Update on the Gestational Effects of Maternal Hyperthermia," *Birth Defects Research 112*: 12 (2020): 943–952.

72 *And exposure to heat in the second and third trimesters*: Johnathan C. K. Wells, "Thermal Environment and Human Birth Weight," *Journal of Theoretical Biology 214*: 3 (2002): 413–425.

72 *Low-birth-weight babies grow into adults*: Keith M. Godfrey and David J. P. Barker, "Fetal Programming and Adult Health," *Public Health Nutrition 4*: 2b (2001): 611–624; Gary C. Curhan, Walter C. Willett, Eric B. Rimm et al., "Birth Weight and Adult Hypertension, Diabetes Mellitus, and Obesity in US Men," *Circulation 94*: 12 (1996): 3246–3250; Richard J. Silverwood, Mary Pierce, Rebecca Hardy et al., "Low Birth Weight, Later Renal Function, and the Roles of Adulthood Blood Pressure, Diabetes, and Obesity in a British Birth Cohort," *Kidney International 84*: 6 (2013): 1262–1270; Caitlin J. Smith and Kelli K. Ryckman, "Epigenetic and Developmental Influences on the Risk of Obesity, Diabetes, and Metabolic Syndrome," *Diabetes, Metabolic Syndrome and Obesity: Targets and Therapy 8* (2015): 295–302.

72 *The heat risks to pregnant women*: Bekkar et al., "Association of Air Pollution and Heat Exposure with Preterm Birth, Low Birth Weight, and Stillbirth in the US: A Systematic Review."

73 *Children with asthma sometimes cough and wheeze*: Zhiwei Xu, Cunrui Huang, Wenbiao Hu et al., "Extreme Temperatures and Emergency Department Admissions for Childhood Asthma in Brisbane, Australia," *Occupational and Environmental Medicine 70*: 10 (2013): 730–735.

73 *Diabetic children have higher blood sugars*: Lisanne L. Blauw, N. Ahmad Aziz, Martijn R. Tannemaat et al., "Diabetes Incidence and Glucose Intolerance Prevalence Increase with Higher Outdoor Temperature," *BMJ Open Diabetes Research and Care 5* (2017): e000317; Zhiwei Xu, Shilu Tong, Jian Cheng et al., "Heatwaves and Diabetes in Brisbane, Australia: A Population-Based Retrospective Cohort Study," *International Journal of Epidemiology 48*: 4 (2019): 1091–1100; Yiwen Zhang, Rongbin Xu, Tingting Ye et al., "Heat Exposure and Hospitalization for Epileptic

Seizures: A Nationwide Case-Crossover Study in Brazil," *Urban Climate 49* (2023): 101497; Medine I. Gulcebi, Emanuele Bartolini, Omay Lee et al., "Climate Change and Epilepsy: Insights from Clinical and Basic Science Studies," *Epilepsy & Behavior 116* (2021): 107791.

73 *Kidney injuries increase because of dehydration*: Xiao-Yu Wang, Adrian Barnett, Yu-Ming Guo et al., "Increased Risk of Emergency Hospital Admissions for Children with Renal Diseases During Heatwaves in Brisbane, Australia," *World Journal of Pediatrics 10* (2014): 330–335; O. Bar-Or, John A. Hay, D. S. Ward et al., "Voluntary Dehydration and Heat Intolerance in Cystic Fibrosis," *Lancet 339*: 8795 (1992): 696–699.

73 *For all these reasons, pediatric emergency room visits*: Aaron S. Bernstein, Shengzhi Sun, Kate R. Weinberger et al., "Warm Season and Emergency Department Visits to U.S. Children's Hospitals," *Environmental Health Perspectives 130*: 1 (2022): 017001.

Heat and Human Behavior

73 *A growing field of research*: Andreas Miles-Novelo and Craig A. Anderson, "Climate Change and Psychology: Effects of Rapid Global Warming on Violence and Aggression," *Current Climate Change Reports 5* (2019): 36–46.

74 *Experiments done by psychologists*: Robert A. Baron and Paul A. Bell, "Aggression and Heat: The Influence of Ambient Temperature, Negative Affect, and a Cooling Drink on Physical Aggression," *Journal of Personality and Social Psychology 33*: 3 (1976): 245–255.

74 *In hot weather, drivers are more likely to honk*: Douglas T. Kenrick and Steven W. MacFarlane, "Ambient Temperature and Horn Honking: A Field Study of the Heat/Aggression Relationship," *Environment and Behavior 18*: 2 (1986): 179–191.

74 *Baseball pitchers hit batters more often*: Alan S. Reifman, Richard P. Larrick, and Steven Fein, "Temper and Temperature on the Diamond: The Heat-Aggression Relationship in Major League Baseball," *Personality and Social Psychology Bulletin 17*: 5 (1991): 580–585.

74 *Violent crime rates go up*: Jan Tiihonen, Pirjo Halonen, Laura Tiihonen et al., "The Association of Ambient Temperature and Violent Crime," *Scientific Reports 7*: 1 (2017): 6543; Rahini Mahendran, Rongbin Xu, Shanshan Li, and Yuming Guo, "Interpersonal Violence Associated with Hot Weather," *Lancet Planetary Health 5*: 9 (2021): e571–e572.

74 *Adult depression and other mental health problems*: Jingwen Liu, Blesson M. Varghese, Alana Hansen et al., "Is There an Association Between Hot Weather and Poor Mental Health Outcomes? A Systematic Review and Meta-analysis," *Environment International 153* (2021): 106533; Nick Obradovich, Robyn Migliorini, Martin P. Paulus, and Iyad Rahwan, "Empirical Evidence of Mental Health Risks Posed by Climate Change," *Proceedings of the National Academy of Sciences 115*: 43 (2018): 10953–10958; Amruta Nori-Sarma, Shengzhi Sun, Yuantong Sun et al., "Association Between Ambient Heat and Risk of Emergency Department Visits for Mental Health Among US Adults, 2010 to 2019," *JAMA Psychiatry 79*: 4 (2022): 341–349.

74 *Witnessing violence between parents*: Gayla Margolin, "Effects of Domestic Violence on Children," in Penelope K. Trickett and Cynthia J. Schellenbach, eds., *Violence Against Children in the Family and the Community* (Washington, DC: American Psychologi-

cal Association, 1998), pp. 57–101; US Department of Health and Human Services (HHS) Administration for Children and Families (ACF), Administration on Children, Youth, and Families (ACYF), Children's Bureau, Child Welfare Information Gateway, *State Statutes: Child Witnesses to Domestic Violence* (2021), p. 2, *childwelfare.gov*, accessed June 30, 2023, https://www.childwelfare.gov/pubPDFs/witnessdv.pdf.

74 *Over a half million children are abused or neglected*: HHS, ACF, ACYF, Children's Bureau, *Child Maltreatment 2021* (2023), p. 34, *acf.hhs.gov*, accessed July 3, 2023, https://www.acf.hhs.gov/cb/data-research/child-maltreatment.

74 *In 2019, a group of Oklahoma City physicians*: Blake C. Gruenberg, Ryan D. Brown, Michael P. Anderson, and Amanda L. Bogie, "The Link Between Temperature and Child Abuse," *Trauma 4* (2019): 1–5.

74 *And an analysis of child protective services records*: Jessica Pac, "Hot Tempered: New Evidence on Temperature and Child Abuse and Neglect," PhD dissertation for the Columbia University School of Social Work (2019), accessed July 5, 2023, https://appam.confex.com/data/extendedabstract/appam/2019/Paper_32638_extended abstract_1992_0.pdf.

74 *Babies and toddlers are always most at risk*: Mary F. Evans, Ludovica Gazze, and Jessamyn Schaller, "Temperature and Maltreatment of Young Children," National Bureau of Economic Research, Working Paper 31522 (2023), *nber.org*, accessed December 16, 2023, https://www.nber.org/papers/w31521.

74 *Across the globe*: Marshall Burke, Solomon M. Hsiang, and Edward Miguel, "Climate and Conflict," *Annual Review of Economics 7*: 1 (2015): 577–617.

74 *Like their parents, kids get irritated when hot*: Anna Malmquist, Tora Lundgren, Mattias Hjerpe et al., "Vulnerability and Adaptation to Heat Waves in Preschools: Experiences, Impacts and Responses by Unit Heads, Educators and Parents," *Climate Risk Management 31* (2021): 100271.

Babies and Football Players

75 *On a summer day, an infant left in a closed car*: Jennifer K. Vanos, Ariane Middel, Michelle N. Poletti, and Nancy J. Selover, "Evaluating the Impact of Solar Radiation on Pediatric Heat Balance Within Enclosed, Hot Vehicles," *Temperature 5*: 3 (2018): 276–292.

75 Vehicular heatstroke *has killed*: "Heatstroke Deaths of Children in Vehicles," *noheatstroke.org*, accessed September 3, 2023, https://www.noheatstroke.org.

75 *In light of this growing toll*: Eric Levenson, "The Bipartisan Infrastructure Bill Takes on Heat Car Deaths, but Advocates Say the Effort Isn't Enough," *CNN*, August 12, 2021, accessed July 3, 2023, https://www.cnn.com/2021/08/12/us/hot-car-deaths-infrastructure-bill/index.html.

75 *The major car manufacturers*: Keith Laing, "Carmakers Agree to Add Rear-Seat Reminders to Prevent Deaths in Hot Cars," *Detroit News*, September 4, 2019, accessed July 3, 2023, https://www.detroitnews.com/story/business/autos/2019/09/04/carmakers-agree-add-rear-seat-child-reminders-2025/2201973001/#.

75 *Some automakers are going further*: Kids and Car Safety, "Examples of Available Technology to Prevent Hot Car Deaths," *kidsandcars.org*, September 20, 2019, accessed July 3, 2023, https://www.kidsandcars.org/news/post/examples-of-available-technology-to

-prevent-hot-car-deaths; Joey Klender, "Tesla Safety Tech Takes Giant Step with FCC Approval for Wave Sensor," *teslarati.com*, April 23, 2021, accessed July 3, 2023, https://www.teslarati.com/tesla-driver-monitoring-system-fcc-filing-granted-child-safety/.

75 *Over 9,200 American high school athletes*: Ellen E. Yard, Julie Gilchrist, Tadesse Haileyesus et al., "Heat Illness Among High School Athletes—United States, 2005–2009," *Journal of Safety Research 41*: 6 (2010): 471–474.

76 *Their fatal cases tripled*: Andrew J. Grundstein, Craig Ramseyer, Fang Zhao et al., "A Retrospective Analysis of American Football Hyperthermia Deaths in the United States," *International Journal of Biometeorology 56* (2012): 11–20.

76 *The climate crisis is largely to blame*: James Bruggers, "'This Was Preventable': Football Heat Deaths and the Rising Temperature," *insideclimatenews.org*, July 20, 2018, accessed July 3, 2023, https://insideclimatenews.org/news/20072018/high-school-football-practice-heat-stroke-exhaustion-deaths-state-rankings-health-safety/.

76 *Many players who die from heatstroke are also obese*: Andrew J. Grundstein et al., "A Retrospective Analysis of American Football Hyperthermia Deaths in the United States."

Valley of the Sun

76 *The news crew from ABC15 caught it from overhead*: Max Walker, "Phoenix Police Identify 12-Year-Old Boy Who Died After Being Airlifted in Mountain Rescue," *ABC15*, July 22, 2016, accessed June 10, 2017, https://www.abc15.com/news/region-phoenix-metro/north-phoenix/phoenix-firefighters-involved-in-mountain-rescue-of-12-year-old-boy.

77 *They ranged from a nineteen-year-old who had moved*: Patricia Madej, "Former North Bend Woman Dies Hiking in Arizona Heat," *Seattle Times*, June 21, 2016, accessed July 2, 2023, https://www.seattletimes.com/seattle-news/former-north-bend-woman-dies-hiking-in-arizona-heat/; "In Memoriam: Stefan Günster, 1960–2016," *optica.org*, June 19, 2016, accessed July 2, 2023, https://www.optica.org/en-us/about/newsroom/obituaries/2016/stefan_gunster/; Sydney Greene, "25-Year-Old Phoenix Man Dies on Peralta Trail Near Gold Canyon," *Arizona Republic*, June 18, 2016, accessed July 2, 2023, https://www.azcentral.com/story/news/local/pinal-breaking/2016/06/18/pinal-county-25-year-old-man-dies-peralta-trail-heat/86104828/; Yihyun Jeong and Adrian Hedden, "Woman Dead Following Phoenix Mountain Rescue," *Arizona Republic*, June 19, 2016, accessed July 2, 2023, https://www.azcentral.com/story/news/local/phoenix-breaking/2016/06/19/phoenix-fire-crews-give-woman-cpr-during-mountain-rescue/86113792/; Leigh Garner, "Family of Alabama Woman Who Died Warns Others About Heat-Related Dangers," *CBS42*, July 21, 2016, accessed July 2, 2023, https://www.cbs42.com/news/family-of-alabama-woman-who-died-warns-others-about-heat-related-dangers/.

77 *Mostly elderly and alone*: Iverson et al., "Heat-Associated Mortality in a Hot Climate: Maricopa County, Arizona, 2006–2016," pp. 11, 15, 17.

77 *They were the kind of people*: Erik Klinenberg, *Heat Wave: A Social Autopsy of Disaster in Chicago*, 2d ed. (Chicago: University of Chicago Press, 2015), p. xxi.

78 *I asked Nancy Selover*: Nancy Selover, personal communication, June 20, 2017.

78 *And though television reporters did not call it to the public's attention*: NOAA, NWS,

"Phoenix, AZ, Climate and Past Weather, NOWData—NOAA Online Weather Data: Daily Data for a Month: July 2016," *weather.gov*, accessed March 3, 2023, https://www.weather.gov/wrh/Climate?wfo=psr.

78 *Multiple agencies and groups are pushing*: Sarah Kaplan and Cassidy Araiza, "How America's Hottest City Will Survive Climate Change," *Washington Post*, July 8, 2020, accessed March 3, 2023, https://www.washingtonpost.com/graphics/2020/climate-solutions/phoenix-climate-change-heat/; Nina Lakhani, "America's Hottest City Is Nearly Unlivable in Summer. Can Cooling Technologies Save It?," *Guardian*, January 27, 2022, accessed March 3, 2023, https://www.theguardian.com/us-news/2022/jan/27/phoenix-arizona-hottest-city-cooling-technologies; Maggie Messerschmidt, Melissa Guardaro, Jessica White et al., *Heat Action Planning Guide for Neighborhoods of Greater Phoenix* (The Nature Conservancy, Maricopa County Public Health, Arizona State University, and others, 2019), accessed March 3, 2023, https://keep.lib.asu.edu/items/141415.

78 *Sadly, their use was curtailed by COVID-19*: Melissa Guardaro, Matthew Roach, David Hondula et al., *Operating Cooling Centers in Arizona Under COVID-19 and Record Heat Conditions: Lessons Learned from Summer 2020* (Arizona Heat Resilience Work Group, 2021), accessed March 3, 2023, https://www.azdhs.gov/documents/preparedness/epidemiology-disease-control/extreme-weather/pubs/gfl-cooling-center-report-2020.pdf.

78 *Dave Hondula, a heat-health researcher*: David Hondula, personal communication, June 7, 2017.

79 *That's because for now, most air-conditioning is powered*: EPA, "Greenhouse Gas (GHG) Emissions and Removals," *epa.gov*, last updated June 21, 2023, accessed July 1, 2023, https://www.epa.gov/ghgemissions.

79 *Thankfully, scientists have developed*: Lisa Friedman and Coral Davenport, "Senate Ratifies Pact to Curb a Broad Category of Potent Greenhouse Gases," *New York Times*, September 21, 2022, accessed July 1, 2023, https://www.nytimes.com/2022/09/21/climate/hydrofluorocarbons-hfcs-kigali-amendment.html.

79 *Yet the year after I traced Cody's steps*: Ryan Randazzo, "APS Parent Company Spent $37.9M Fighting Clean Energy Measure," *Arizona Republic*, January 17, 2019, accessed July 2, 2023, https://www.azcentral.com/story/news/politics/arizona/2019/01/17/pinnacle-west-spent-38-million-fight-arizonas-prop-127-clean-energy-measure/2595711002/.

79 *That reality led Phoenix*: City of Phoenix, "Phoenix Joins C40 Cities," January 21, 2020, *phoenix.gov*, accessed July 2, 2023, https://www.phoenix.gov/newsroom/environmental-programs/838.

80 *Every year hundreds of heat-sick hikers*: Amaris Encinas, "As Heat Settles In, More Hikers Are Being Rescued from Phoenix Trails. Here's How to Stay Safe," *Arizona Republic*, March 31, 2022, accessed July 2, 2023, https://www.azcentral.com/story/news/local/phoenix/2022/03/31/phoenix-hikers-should-know-these-guidelines-before-hitting-any-trails/7234219001/.

80 *Just three weeks before that drive*: Fernanda Santos, "A Proposed Hiking Ban in Phoenix Draws Outrage," *New York Times*, July 1, 2016, accessed July 2, 2023, https://www.nytimes.com/2016/07/02/us/a-proposed-hiking-ban-in-phoenix-draws-heat.html.

Cody's Voice

82 *With each successive year since Cody Flom walked up the mountain*: Maricopa County Public Health, "Heat Reports: Yearly Mortality Reports," *maricopa.gov*, accessed March 3, 2023, https://www.maricopa.gov/1858/Heat-Surveillance.

82 *At least 579 people were killed*: Maricopa County Public Health, "2023 Weekly Heat Report: November 7, 2023," *maricopa.gov*, accessed December 18, 2023, https://www.maricopa.gov/ArchiveCenter/ViewFile/Item/5734.

82 *By late 2021 the Phoenix parks department decided*: Holly Bock, "Phoenix Makes Ban on Hiking in Extreme Heat Permanent," *AZ Family, Phoenix Arizona News, 3TV/CBS5*, October 28, 2021, accessed July 8, 2023, https://www.youtube.com/watch?v=HdUfDEmDN4c.

82 *A dozen firefighters had developed heat illness*: Ali Vetnar, "Phoenix Firefighters Hospitalized After Mountain Rescues in Extreme Heat," *KTAR*, June 25, 2021, accessed July 5, 2023, https://ktar.com/story/4520893/phoenix-firefighters-hospitalized-after-mountain-rescues-in-extreme-heat/; Ray Stern, "At Firefighters' Request, Phoenix Will Close Popular Trails in Extreme Heat," *Phoenix New Times*, July 14, 2021, accessed July 5, 2023, https://www.phoenixnewtimes.com/news/phoenix-trails-closed-for-excessive-heat-piestewa-peak-echo-canyon-camelback-11588975.

82 *A stunning spike in global temperatures*: Hausfather, "State of the Climate: 2023 Smashes Records for Surface Temperatures and Ocean Heat."

82 *Phoenix's summer would be its hottest ever*: Kylee Cruz, "2023 Expected to Be Hottest Summer in Phoenix's History," *azfamily.com*, accessed September 3, 2023, https://www.azfamily.com/2023/08/29/2023-expected-be-hottest-summer-phoenixs-history/; NOAA, NWS, "Phoenix, AZ, Climate and Past Weather, NOWData—NOAA Online Weather Data: Daily Data for a Month: June, July, August 2023," *weather.gov*, accessed December 9, 2023, https://www.weather.gov/wrh/Climate?wfo=psr.

83 *Though Arizona is one of the sunniest places on Earth*: US Energy Information Administration, "Arizona State Energy Profile: Arizona Quick Facts," *eia.gov*, accessed September 10, 2023, https://www.eia.gov/state/print.php?sid=AZ.

Chapter Three: Home

Turning Up the Dial

87 *I had watched a video shot here*: Jeff Ehling, "Boat Ride Reveals Flood Devastation in Meyerland Area," *ABC13 Eyewitness News*, August 28, 2017, accessed March 10, 2023, https://abc13.com/houston-meyerland-water-rescue-tropical-storm-harvey/2350175/.

87 *"As Harvey's rains unfolded"*: Jason Samenow, "Harvey Is a 1,000-Year Flood Event Unprecedented in Scale," *Washington Post*, August 31, 2017, accessed March 10, 2023, https://www.washingtonpost.com/news/capital-weather-gang/wp/2017/08/31/harvey-is-a-1000-year-flood-event-unprecedented-in-scale.

88 *As the hurricane spun in place*: NOAA, NWS, "Tropical Weather: Hurricane Harvey, August 17–September 1, 2017," *weather.gov*, accessed March 10, 2023, https://www.weather.gov/lch/2017harvey.

88 *More than 60,000 people in Harris County had to be rescued*: City of Houston, Texas, *Hurricane Harvey Recovery: A Progress Report* (2019), p. 4, *houstontx.gov*, accessed March 10, 2023, https://www.houstontx.gov/postharvey/public/documents/11.28.2018_progress_report_updated.pdf.

88 *Nearly half of all housing*: Keaton Fox, "Updated Analysis Shows Nearly Half of Houston Homes Damaged in Harvey," *ABC13 Eyewitness News*, October 11, 2018, accessed March 10, 2023, https://abc13.com/half-of-houstons-homes-were-damaged-in-hurricane-harvey/4461534/; David McClendon, assistant director, Center for Social Measurement and Evaluation, Children at Risk, personal communication, March 21, 2018. Per Dr. McClendon, 600,000 children lived in households that applied for FEMA assistance in Texas counties affected by Harvey.

88 *That year, 2017, began with news*: NASA, "NASA, NOAA Data Show 2016 Warmest Year on Record Globally"; NOAA, National Centers for Environmental Information (NCEI), "Billion-Dollar Weather and Climate Disasters," *ncei.noaa.gov*, accessed March 10, 2023, https://www.ncei.noaa.gov/access/billions/.

89 *Globally, natural disasters have increased fivefold*: World Meteorological Organization, "Weather-Related Disasters Increase over Past 50 Years, Causing More Damage but Fewer Deaths," *wmo.int*, August 31, 2021, accessed March 10, 2023, https://public.wmo.int/en/media/press-release/weather-related-disasters-increase-over-past-50-years-causing-more-damage-fewer.

89 *Extreme weather events*: Reza Marsooli, Ning Lin, Kerry Emanuel, and Kairui Feng, "Climate Change Exacerbates Hurricane Flood Hazards Along US Atlantic and Gulf Coasts in Spatially Varying Patterns," *Nature Communications 10*: 1 (2019): 3785; Angela Colbert, "A Force of Nature: Hurricanes in a Changing Climate," *climate.nasa.gov*, June 1, 2022, accessed March 10, 2023, https://climate.nasa.gov/news/3184/a-force-of-nature-hurricanes-in-a-changing-climate/.

Young Minds in a Chaotic World

90 *Though we often associate PTSD with soldiers*: Jessica Hamblen and Erin Barnett, "PTSD in Children and Adolescents," *National Center for PTSD* (2016), *ptsd.va.gov*, accessed July 6, 2023, https://www.ptsd.va.gov/professional/treat/specific/ptsd_child_teens.asp.

90 *Natural disasters always leave an epidemic of PTSD*: Yuval Neria, Arijit Nandi, and Sandro Galea, "Post-Traumatic Stress Disorder Following Disasters: A Systematic Review," *Psychological Medicine 38*: 4 (2008): 467–480.

91 *Hurricane Katrina displaced more than a million people*: US Department of Housing and Urban Development, "A Look Back at Hurricane Katrina," *huduser.gov*, September 21, 2021, accessed March 10, 2023, https://www.huduser.gov/portal/pdredge/pdr-edge-frm-asst-sec-092121.html.

91 *Later that year, I was not surprised*: David M. Abramson, Yoon Soo Park, Tasha Stehling-Ariza, and Irwin Redlener, "Children as Bellwethers of Recovery: Dysfunctional Systems and the Effects of Parents, Households, and Neighborhoods on Serious Emotional Disturbance in Children After Hurricane Katrina," *Disaster Medicine and Public Health Preparedness 4*: S1 (2010): S17–S27.

91 *Other researchers would find*: Cynthia L. Rowe, Annette M. La Greca, and Anders Alexandersson, "Family and Individual Factors Associated with Substance Involvement and PTS Symptoms Among Adolescents in Greater New Orleans After Hurricane Katrina," *Journal of Consulting and Clinical Psychology 78*: 6 (2010): 806–817.

91 *Like Nia, hundreds of thousands of New Orleans children*: Katy Reckdahl, "The Lost Children of Katrina," *Atlantic*, April 2, 2015, accessed July 8, 2023, https://www.theatlantic.com/education/archive/2015/04/the-lost-children-of-katrina/389345/.

92 *She was one of many New Orleans children*: Barbara Rath, Elizabeth A. Young, Amy Harris et al., "Adverse Respiratory Symptoms and Environmental Exposures Among Children and Adolescents Following Hurricane Katrina," *Public Health Reports 126*: 6 (2011): 853–860.

92 *Harvey alone was unprecedented*: NOAA, NWS, "Tropical Weather: Hurricane Harvey."

93 *We know that its magnitude was due to climate change*: Henry Fountain, "Scientists Link Hurricane Harvey's Record Rainfall to Climate Change," *New York Times*, December 13, 2017, accessed March 10, 2023, https://www.nytimes.com/2017/12/13/climate/hurricane-harvey-climate-change.html.

93 *Shani, like Lucas, was just five years old*: Debra Kahn and Anne C. Mulkern, "Scientists See Climate Change in California's Wildfires," *scientificamerican.com*, October 12, 2017, accessed March 11, 2023, https://www.scientificamerican.com/article/scientists-see-climate-change-in-californias-wildfires/.

94 *Here, in the early hours of October 9, 2017*: CalFire, "Tubbs Fire (Central LNU Complex)," *fire.ca.gov*, October 25, 2019, accessed March 11, 2023, https://www.fire.ca.gov/incidents/2017/10/8/tubbs-fire-central-lnu-complex/.

95 *The local news celebrated Coffey Park's revival*: Adia White, "A Third of Homes Lost in Tubbs Fire Now Being Rebuilt," *KQED*, April 20, 2019, accessed March 11, 2023, https://www.kqed.org/news/11741498/a-third-of-homes-lost-in-2017-tubbs-fire-now-under-construction.

95 *But within weeks the area would evacuate again*: Colleen Shalby, "These Santa Rosa Residents Just Rebuilt Their Homes. Then Came the Kincade Fire," *Los Angeles Times*, October 28, 2019, accessed March 11, 2023, https://www.latimes.com/california/story/2019-10-28/kincade-fire-sonoma-coffey-park-tubbs-evacuation-rebuild.

95 Eighteen *major wildfires*: John Blanchard and J. D. Morris, "The Fires That Have Scarred Wine Country," *San Francisco Chronicle*, September 29, 2020, accessed March 11, 2023, https://www.sfchronicle.com/projects/2020/wine-country-fires-map/.

95 *In the fall of 2020, they looked up at skies*: Kari Paul, "'Good Morning, Hell': Californians Awake to Apocalyptic Skies as Wildfires Rage," *Guardian*, September 9, 2020, accessed March 11, 2023, https://www.theguardian.com/us-news/2020/sep/09/orange-sky-california-fires-smoke-san-francisco.

95 *Lucas's situation is similar*: Associated Press, "Why Does Houston Flood So Often and So Heavily?," *NBC News*, August 27, 2017, accessed March 11, 2023, https://www.nbcnews.com/storyline/hurricane-harvey/why-does-houston-flood-so-often-so-heavily-n796446.

The Children of Hurricane Harvey

96 *When I met with Dr. Kaplow*: Carlos Ballesteros, "Hurricane Harvey Victims: More Than 20,000 Children in Houston Are Homeless, Report Shows," *Newsweek*, November 26, 2017, accessed March 11, 2023, https://www.newsweek.com/hurricane-harvey-victims-homeless-fema-722640.

96 *The households of over a half million children had been damaged or destroyed*: David McClendon, personal communication.

96 *At that point, she told me*: Julie Kaplow, personal communication, January 24, 2018.

97 *The answer has been pieced together from studies*: David M. Abramson et al., "Children as Bellwethers of Recovery: Dysfunctional Systems and the Effects of Parents, Households, and Neighborhoods on Serious Emotional Disturbance in Children After Hurricane Katrina"; Jon A. Shaw, Brooks Applegate, and Caryn Schorr, "Twenty-One-Month Follow-up Study of School-Age Children Exposed to Hurricane Andrew," *Journal of the American Academy of Child & Adolescent Psychiatry 35*: 3 (1996): 359–364; David M. Abramson, Donna Van Alst, Alexis Merdjanoff et al., *The Hurricane Sandy Person Report: Disaster Exposure, Health Impacts, Economic Burden, and Social Well-being,* Sandy Child and Family Health Study, Rutgers University School of Social Work, New York University College of Global Public Health, Columbia University National Center for Disaster Preparedness, Colorado State University Center for Disaster and Risk Analysis, Briefing Report No. 2 (2015), *academiccommons.columbia.edu*, accessed March 16, 2023, https://academiccommons.columbia.edu/doi/10.7916/D8ST7P3Q; Matthew R. G. Brown, Vincent Agyapong, Andrew J. Greenshaw et al., "Significant PTSD and Other Mental Health Effects Present 18 Months After the Fort McMurray Wildfire: Findings from 3,070 Grades 7–12 Students," *Frontiers in Psychiatry 10* (2019): 623.

97 *In 2019, the year after I met with her*: Cody G. Dodd, Ryan M. Hill, Benjamin Oosterhoff et al., "The Hurricane Exposure, Adversity, and Recovery Tool (HEART): Developing and Validating a Risk Screening Instrument for Youth Exposed to Hurricane Harvey," *Journal of Family Strengths 19*: 1 (2019): Article 5, 1–26.

98 *Adults under severe stress*: Sandra M. Stith, Ting Liu, L. Christopher Davies et al., "Risk Factors in Child Maltreatment: A Meta-analytic Review of the Literature," *Aggression and Violent Behavior 14*: 1 (2009): 13–29.

99 *If they are severely distressed and struggling*: Children's Environmental Health Initiative (CEHI), "Texas Flood Registry 2020 Report," *harveyregistry.rice.edu*, accessed March 12, 2023, https://harveyregistry.rice.edu/third_party/HHR_2020_Report_English.pdf.

99 *Early results showed that these efforts were working*: Julie Kaplow, personal communication.

99 *By Harvey's two-year anniversary*: "Trauma and Grief Reactions Among Children More Apparent Two Years After Hurricane Harvey," *Huntsville Item*, August 12, 2019, accessed March 12, 2023, https://www.itemonline.com/news/trauma-and-grief-reactions-among-children-more-apparent-two-years-after-hurricane-harvey/article_c8ddff12-bd17-11e9-8dc7-37498d5ab8a6.htm.

99 *Harvey was just one disaster out of many*: NOAA, NCEI, "US Billion-Dollar Weather and Climate Disasters."

99 *The trend shows no sign of slowing*: Ibid., accessed December 15, 2023.

100 *Fortunately, the Harvey program's approach*: National Child Traumatic Stress Network (NCTSN), "About Us," *nctsn.org*, accessed July 2, 2023, https://www.nctsn.org/about-us.

100 *These include trauma- and grief-informed assessment tools*: NCTSN, "All NCTSN Resources," *nctsn.org*, accessed March 12, 2023, https://www.nctsn.org/resources/all-nctsn-resources.

Children's Brains Are Different

102 *This phenomenon, known as* neuroplasticity: Bryan Kolb and Robbin Gibb, "Brain Plasticity and Behaviour in the Developing Brain," *Journal of the Canadian Academy of Child and Adolescent Psychiatry 20*: 4 (2011): 265–276.

102 *When Lucas was born, his brain already contained*: The research supporting the "Children's Brains Are Different" section is summarized by: Lise Eliot, *What's Going on In There? How the Brain and Mind Develop in the First Five Years of Life* (New York: Bantam Books, 1999); Daniel J. Siegel, "Toward an Interpersonal Neurobiology of the Developing Mind: Attachment Relationships, 'Mindsight,' and Neural Integration," *Infant Mental Health Journal 22*: 1–2 (2001): 67–94; National Scientific Council on the Developing Child, *Young Children Develop in an Environment of Relationships: Working Paper 1* (2004), and *Children's Emotional Development Is Built into the Architecture of Their Brains: Working Paper 2* (2004), *developingchild.harvard.edu*, both accessed March 20, 2023, https://developingchild.harvard.edu/resourcecategory/reports-and-working-papers/; Louis Cozolino, *The Neuroscience of Human Relationships, Attachment and the Developing Social Brain*, 2d ed. (New York: W. W. Norton & Company, 2014).

Toxic Stress

105 *But Sophia was likely exposed in her early life*: The research supporting the "Toxic Stress" section is summarized by: National Scientific Council on the Developing Child, *Excessive Stress Disrupts the Architecture of the Developing Brain: Working Paper 3*, updated edition (2005/2014), *developingchild.harvard.edu*, accessed March 20, 2023, https://developingchild.harvard.edu/resourcecategory/reports-and-working-papers/.

108 *But as one group of researchers put it*: Ibid., p. 4.

Unnatural Disasters

109 *That changed dramatically on April 14, 1969*: NOAA, NASA, "The World According to Weather Satellites," *nasa.gov*, accessed March 17, 2023, https://storymaps.arcgis.com/stories/3ed6b70ffa80447aac3fcb1d3378884a; Gerald L. Wick, "Nimbus Weather Satellites: Remote Sounding of the Atmosphere," *Science 172*: 3989 (1971): 1222–1223.

109 *For the first time*: Although the TIROS-1, launched in 1960, was officially NASA's

first weather satellite, it was little more than a camera. The Nimbus 3 included instruments that could measure temperature at different altitudes in the atmosphere, air pressure, solar ultraviolet radiation, the ozone layer, and sea ice. It enabled meteorologists to conduct analyses that had not previously been possible and launched modern weather forecasting.

109 *As the first real weather satellite*: Hannah Ritchie, Pablo Rosado, and Max Roser, "Natural Disasters," *ourworldindata.org*, accessed March 16, 2023, https://ourworld indata.org/natural-disasters.

109 *Nowhere did this feat mean more*: NOAA, NASA, "The World According to Weather Satellites"; Erik Larson, *Isaac's Storm: A Man, a Time, and the Deadliest Hurricane in History* (Visalia, CA: Vintage Press, 2000), p. 16.

110 *Between 6,000 and 12,000 people*: Srikanto H. Paul, Hatim O. Sharif, and Abigail M. Crawford, "Fatalities Caused by Hydrometeorological Disasters in Texas," *Geosciences 8*: 5 (2018): 186.

110 This event is unprecedented & all impacts are unknown: NWS, Twitter post, August 27, 2017, 8:44 a.m., *twitter.com/NWS*, accessed March 16, 2023, https://twitter .com/NWS/status/901832717070983169.

110 *The new satellite could tell us the types of gases*: Though the Nimbus 3 did not measure atmospheric gas concentrations directly, the data it collected allowed for these concentrations to be calculated, as explained by these sources: Lieven Clarisse, Yasmina R'Honi, Pierre-François Coheur et al., "Thermal Infrared Nadir Observations of 24 Atmospheric Gases," *Geophysical Research Letters 38*: 10 (2011): L10802; Allied Research Associates, Inc., Prepared for NASA, Goddard Space Flight Center, Greenbelt, MD, *The Best of Nimbus*, Contract No. NAS 5-10343 (1971), p. 98, accessed July 5, 2023, https://core.ac.uk/download/pdf/80643021.pdf.

110 *These oil giants would be silent when*: Fountain, "Scientists Link Hurricane Harvey's Record Rainfall to Climate Change."

111 *Texas Children's itself was not spared*: Michael Braun, Helen Currier, and staff of Renal Center at Texas Children's Hospital, personal communication, January 22, 2018.

111 *Although it has changed its rhetoric*: Mei Li, Gregory Trencher, and Jusen Asuka, "The Clean Energy Claims of BP, Chevron, ExxonMobil and Shell: A Mismatch Between Discourse, Actions and Investments." *PloS One 17*: 2 (2022): e0263596; Chris McGreal, "How a Powerful US Lobby Group Helps Big Oil to Block Climate Action," *Guardian*, July 18, 2021, accessed March 16, 2023, https://www.theguardian. com/environment/2021/jul/19/big-oil-climate-crisis-lobby-group-api.

Love and Water

113 *When the cyclone struck, ninety-three children and ten nuns*: Larson, *Isaac's Storm*, pp. 212–213.

113 *I knew that the ocean was 2 feet higher*: NOAA, "Relative Sea Level Trend, Tides and Currents, 8771450 Galveston Pier 21, Texas," *tidesandcurrents.noaa.gov*, accessed March 16, 2023, https://tidesandcurrents.noaa.gov/sltrends/sltrends_station.shtm l?id=8771450.

114 *Only four months after the Galveston hurricane*: Judith Walker, Ellen Walker Rienstra,

Jo Ann Stiles et al., *Giant Under the Hill: A History of the Spindletop Oil Discovery at Beaumont, Texas, in 1901* (Austin, TX: Texas State Historical Association, 2002).

114 *Hurricanes are becoming stronger*: Michael F. Wehner and James P. Kossin, "The Growing Inadequacy of an Open-Ended Saffir-Simpson Hurricane Wind Scale in a Warming World," *Proceedings of the National Academy of Sciences 121*: 7 (2024): e2308901121.

114 *And they move on more slowly once they strike*: Emily Atkin, "Hurricane Sally, Mutant Sloth," *heated.world*, September 16, 2020, accessed July 15, 2023, https://heated.world/p/hurricane-sally-mutant-sloth.

114 *Anyone can go on NASA's website*: NASA Earth Observatory, Images: Natural Events, *earthobservatory.nasa.gov*, accessed March 16, 2023, https://earthobservatory.nasa.gov/images.

114 *Many oil and gas facilities had been closed and damaged by Harvey*: Ben Lefebvre, "Harvey Triggers Spike in Hazardous Chemical Releases," *Politico*, August 29, 2017, accessed March 16, 2023, https://www.politico.com/story/2017/08/29/hurricane-harvey-chemical-danger-242142; Rae Lynn Mitchell, Texas A&M University School of Public Health, "Following Hurricane Harvey, Pollutant Levels in Houston Neighborhood Exceeded Limit for Increased Cancer Risks," *Texas A&M Today*, March 26, 2021, accessed March 16, 2023, https://today.tamu.edu/2021/03/26/following-hurricane-harvey-pollutant-levels-in-houston-neighborhood-exceeded-limit-for-increased-cancer-risks/.

Chapter Four: Small Life

117 *Only a few cases of sexual transmission*: Didier Musso, Claudine Roche, Emelie Robin et al., "Potential Sexual Transmission of Zika Virus," *Emerging Infectious Diseases 21*: 2 (2015): 359–361, erratum in: *Emerging Infectious Diseases 21*: 3 (2015): 552.

118 *And this little disabled girl*: Harriet Alexander, "First Zika-Affected Baby Born in the United States," *Telegraph*, June 1, 2016, accessed September 3, 2023, https://www.telegraph.co.uk/news/2016/06/01/first-zika-affected-baby-born-in-the-united-states/.

A Plague Among Children

118 *Though it sickened hundreds of thousands*: Pan American Health Organization (PAHO), "Zika-Epidemiological Update," *paho.org*, August 25, 2017, accessed July 2, 2023, https://www.paho.org/en/documents/25-august-2017-zika-epidemiological-update-0.

119 *She was the first baby born in the continental US*: Debra Goldschmidt, "Baby with Zika-Related Microcephaly Born at New Jersey Hospital," *CNN*, June 1, 2016, accessed July 2, 2023, https://www.cnn.com/2016/06/01/health/baby-born-microcephaly-new-jersey/index.html.

119 *This devastating anomaly, which had spiked*: Donald G. McNeil Jr., "Zika Virus, a Mosquito-Borne Infection, May Threaten Brazil's Newborns," *New York Times*, December 28, 2015, accessed July 2, 2023, https://www.nytimes.com/2015/12/29/health/zika-virus-brazil-mosquito-brain-damage.html.

119 *The Zika virus hunted young nerve cells*: Patricia P. Garcez, Erick Correia Loiola, Rodrigo Madeiro da Costa et al., "Zika Virus Impairs Growth in Human Neurospheres and Brain Organoids," *Science 352*: 6287 (2016): 816–818.

119 *Why this previously benign virus*: Hongjie Xia, Huanle Luo, Chao Shan et al., "An Evolutionary NS1 Mutation Enhances Zika Virus Evasion of Host Interferon Induction," *Nature Communications 9*: 1 (2018): 414.

119 *We had been assured by public health officials*: James Wilson, "Placing the Zika Virus in a Broader Perspective," *Nevada Today*, April 26, 2016, accessed July 2, 2023, https://www.unr.edu/nevada-today/blogs/2016/zika-virus-in-perspective.

119 *Yet the virus was moving north*: Donald G. McNeil Jr., "Zika Cases in Puerto Rico Are Skyrocketing," *New York Times*, July 30, 2016, accessed July 2, 2023, https://www.nytimes.com/2016/07/31/health/zika-virus-puerto-rico.html; Christine E. Prue, Joseph N. Roth Jr., Amanda Garcia-Williams et al., "Awareness, Beliefs, and Actions Concerning Zika Virus Among Pregnant Women and Community Members—US Virgin Islands, November–December 2016," *Morbidity and Mortality Weekly Report 66*: 34 (2017): 909–913.

Webbed Together

120 *Infections are produced mostly by*: Luis P. Villarreal, "Are Viruses Alive?," *Scientific American 291*: 6 (2004): 100–105.

121 *My distant predecessors didn't know*: John Waller, *The Discovery of the Germ: Twenty Years That Transformed the Way We Think About Disease* (New York: Columbia University Press, 2004).

121 *While childhood mortality in the US has been greatly reduced*: Max Roser, "Mortality in the Past—Around Half Died as Children," *ourworldindata.com*, June 11, 2019, accessed July 3, 2023, https://ourworldindata.org/child-mortality-in-the-past.

121 *Yet escalating floods and droughts*: Claire D. Bourke, James A. Berkley, and Andrew J. Prendergast, "Immune Dysfunction as a Cause and Consequence of Malnutrition," *Trends in Immunology 37*: 6 (2016): 386–398; Emma Carlsson, Anneli Frostell, Johnny Ludvigsson, and Maria Faresjö, "Psychological Stress in Children May Alter the Immune Response," *Journal of Immunology 192*: 5 (2016): 2071–2081.

122 *In the African country of Madagascar*: Doctors Without Borders, "Food Assistance Urgently Needed as Nutrition Crisis Grips Southern Madagascar," *msf.org*, accessed October 18, 2023, https://www.msf.org/drought-leaves-skyrocketing-levels-malnutrition-Madagascar.

122 *And after devastating floods struck Pakistan*: Syed Raza Hassan and Asif Shahzad, "Children, Women Prone to Diseases in Pakistan's Stagnant Flood Waters," *Reuters*, September 16, 2022, accessed January 6, 2024, http://www.reuters.com/world/asia-pacific/disease-spreads-pakistan-flooding-toll-surpasses-1500-2022-09-16/.

122 *Unfortunately, natural disasters aren't the only*: Peter H. Raven and David L. Wagner, "Agricultural Intensification and Climate Change Are Rapidly Decreasing Insect Biodiversity," *Proceedings of the National Academy of Sciences 118*: 2 (2021); Helen N. Fones, Daniel P. Bebber, Thomas M. Chaloner et al., "Threats to Global Food Security from Emerging Fungal and Oomycete Crop Pathogens," *Nature Food 1* (2020): 332–342; Martin C. Hänsel, Jörn O. Schmidt, Martina H. Stiasny et al., "Ocean Warming and Acidification May Drag Down the Commercial Arctic Cod Fishery by 2100," *PLoS*

One 15: 4 (2020): e0231589; Samuel S. Myers, Antonella Zanobetti, Itai Kloog et al., "Increasing CO_2 Threatens Human Nutrition," *Nature 510*: 7503 (2014): 139–142.

122 *Because infants and toddlers need 3 to 4 times*: Perera, *Children's Health and the Peril of Climate Change*, p. 35.

122 *That's why malnutrition not only harms them directly*: UNICEF, "Child Malnutrition," *unicef.org*, May 2023, accessed July 15, 2023, https://data.unicef.org/topic/nutrition/malnutrition/.

122 *Serious infections then increase a child's need for energy*: Cutberto Garza, "Effect of Infection on Energy Requirements of Infants and Children," *Public Health Nutrition 8*: 7A (2005): 1187–1190.

122 *In fact, starvation in young children*: Judd L. Walson and James A. Berkley, "The Impact of Malnutrition on Childhood Infections," *Current Opinion in Infectious Diseases 31*: 3 (2018): 231–236.

122 *Our advances against childhood ailments*: Tatsuo Sakai and Yuh Morimoto, "The History of Infectious Diseases and Medicine," *Pathogens 11*: 10 (2022): 1147.

123 *A 2022 study found that of 375 known human infections*: Camilo Mora, Tristan McKenzie, Isabella M. Gaw et al., "Over Half of Known Human Pathogenic Diseases Can Be Aggravated by Climate Change," *Nature Climate Change 12*: 9 (2022): 869–875.

Some Like It Hot

123 *Some* fungal infections, *for example*: Aditi Gadre, Wendemagegn Enbiale, Louise K. Andersen, and Sarah J. Coates, "The Effects of Climate Change on Fungal Diseases with Cutaneous Manifestations: A Report from the International Society of Dermatology Climate Change Committee," *Journal of Climate Change and Health 6* (2022): 100156; CDC, "Valley Fever (*Coccidioidomycosis*) Statistics," accessed March 17, 2023, https://www.cdc.gov/fungal/diseases/coccidioidomycosis/statistics.html.

123 Water- and food-borne infections: D. Onozuka and M. Hashizume, "Weather Variability and Paediatric Infectious Gastroenteritis," *Epidemiology & Infection 139*: 9 (2011): 1369–1378; Timothy J. Wade, Cynthia J. Lin, Jyotsna S. Jagai, and Elizabeth D. Hilborn, "Flooding and Emergency Room Visits for Gastrointestinal Illness in Massachusetts: A Case-Crossover Study," *PloS One 9*: 10 (2014): e110474; Pin Wang, Earnest Asare, Virginia E. Pitzer et al., "Associations Between Long-Term Drought and Diarrhea Among Children Under Five in Low- and Middle-Income Countries," *Nature Communications 13*: 1 (2022): 3661; Rebecca Philipsborn, Sharia M. Ahmed, Berry J. Brosi, and Karen Levy, "Climatic Drivers of Diarrheagenic *Escherichia Coli* Incidence: A Systematic Review and Meta-analysis," *Journal of Infectious Diseases 214*: 1 (2016): 6–15; Luma Akil, H. Anwar Ahmad, and Renata S. Reddy, "Effects of Climate Change on Salmonella Infections," *Foodborne Pathogens and Disease 11*: 12 (2014): 974–980; Eirini Christaki, Panagiotis Dimitriou, Katerina Pantavou, and Georgios K. Nikolopoulos, "The Impact of Climate Change on Cholera: A Review on the Global Status and Future Challenges," *Atmosphere 11*: 5 (2020): 449; Maryam Ghazani, Gerard FitzGerald, Wenbiao Hu et al., "Temperature Variability and Gastrointestinal Infections: A Review of Impacts and Future Perspectives," *International Journal of Environmental Research and Public Health 15*: 4 (2018): 766.

123 *Other waterborne bacteria*: Craig Baker-Austin, Joaquin Trinanes, Narjol Gonzalez-Escalona, and Jaime Martinez-Urtaza, "Non-Cholera Vibrios: The Microbial Barometer of Climate Change," *Trends in Microbiology 25*: 1 (2017): 76–84; Justin P. Bandino, Anna Hang, and Scott A. Norton, "The Infectious and Noninfectious Dermatological Consequences of Flooding: A Field Manual for the Responding Provider," *American Journal of Clinical Dermatology 16* (2015): 399–424; Colleen L. Lau, Lee D. Smythe, Scott B. Craig, and Philip Weinstein, "Climate Change, Flooding, Urbanisation and Leptospirosis: Fuelling the Fire?," *Transactions of the Royal Society of Tropical Medicine and Hygiene 104*: 10 (2010): 631–638.

123 *And a brain-invading, freshwater amoeba*: CDC, "*Naegleria fowleri*—Primary Amebic Meningoencephalitis (PAM)—Amebic Encephalitis: Case Reports by State of Exposure," *cdc.gov*, May 3, 2023, accessed June 30, 2023, https://www.cdc.gov/parasites/naegleria/state-map.html.

123 *These infections, which account for up to three-quarters*: Louise H. Taylor, Sophia M. Latham, and Mark E. J. Woolhouse, "Risk Factors for Human Disease Emergence," *Philosophical Transactions of the Royal Society B: Biological Sciences 356*: 1411 (2001): 983–989.

123 *While some zoonotic diseases*: Kate E. Jones, Nikkita G. Patel, Marc A. Levy et al., "Global Trends in Emerging Infectious Diseases," *Nature 451*: 7181 (2008): 990–993.

124 *In fact, today's children are growing up on a planet*: Rosamunde E. A. Almond, Monique Grooten, Diego Juffe Bignoli, and Tanya Petersen, eds., *Living Planet Report 2022—Building a Nature Positive Society* (Gland, Switzerland: World Wildlife Foundation, 2022), p. 4.

124 *As thermometers climb, rainfall patterns change*: Colin J. Carlson, Gregory F. Albery, Cory Merow et al., "Climate Change Increases Cross-Species Viral Transmission Risk," *Nature 607*: 7919 (2022): 555–562.

124 *Our most recent plague, COVID-19, may be an example*: Robert M. Beyer, Andrea Manica, and Camilo Mora, "Shifts in Global Bat Diversity Suggest a Possible Role of Climate Change in the Emergence of SARS-CoV-1 and SARS-CoV-2," *Science of the Total Environment 767* (2021): 145413; Monali C. Rahalkar and Rahul A. Bahulikar, "Lethal Pneumonia Cases in Mojiang Miners (2012) and the Mineshaft Could Provide Important Clues to the Origin of SARS-CoV-2," *Frontiers in Public Health 8* (2020): 638; Peng Zhou, Xing-Lou Yang, Xian-Guang Wang et al., "A Pneumonia Outbreak Associated with a New Coronavirus of Probable Bat Origin," *Nature 579*: 7798 (2020): 270–273.

124 *The mosquito is by far the world's most important vector*: Timothy C. Winegard, *The Mosquito: A Human History of Our Deadliest Predator* (New York: Dutton, 2019), p. 2.

124 *As one infectious disease specialist reminded me*: Amy Vittor, personal communication, February 3, 2023, and "Mosquitos, Ticks, and Climate Change," her presentation at the July 2021 meeting of Florida Clinicians for Climate Action.

The Epic Struggle

126 *A month before his visit, in July 2016*: Jennifer Marshall, Blake Scott, Jennifer Delva et al., "An Evaluation of Florida's Zika Response Using the WHO Health Systems

Framework: Can We Apply These Lessons to COVID-19?", *Maternal & Child Health Journal 24*: 10 (2020): 1212–1223.

126 *Several people with travel-acquired Zika*: "Washoe County Officials Confirm Zika in Pregnant Woman," *Nevada Appeal*, August 26, 2016, accessed July 6, 2023, https://www.nevadaappeal.com/news/2016/aug/26/washoe-county-officials-confirm-zika-in-pregnant-w/.

126 *We didn't know that within a year*: Blake Apgar, "Mosquito Capable of Spreading Zika Found in Southern Nevada," *Las Vegas Review-Journal*, June 1, 2017, accessed July 6, 2023, https://www.reviewjournal.com/local/north-las-vegas/mosquito-capable-of-spreading-zika-found-in-southern-nevada/.

127 *"It was literally the plague among children"*: Noah Webster, *A Brief History of Epidemic and Pestilential Diseases: With the Principal Phenomena of the Physical World, Which Precede and Accompany Them, and Observations Deduced from the Facts Stated, in Two Volumes* (Hartford, CT: Hudson & Goodwin, 1799), p. 233.

127 *A century ago in the US*: Rebecca M. Cunningham, Maureen A. Walton, and Patrick M. Carter, "The Major Causes of Death in Children and Adolescents in the United States," *New England Journal of Medicine 379*: 25 (2018): 2468–2475.

127 *Humanity's success against pediatric infectious disease*: Max Roser, "Our History Is a Battle Against the Microbes: We Lost Terribly Before Science, Public Health, and Vaccines Allowed Us to Protect Ourselves," *ourworldindata.org*, July 20, 2020, accessed July 6, 2023, https://ourworldindata.org/microbes-battle-science-vaccines.

127 *Yet huge challenges remain*: Bernadeta Dadonaite, "What Are Children Dying from and What Can We Do About It?," *ourworldindata.org*, August 9, 2019, accessed July 6, 2023, https://ourworldindata.org/what-are-children-dying-from-and-what-can-we-do-about-it.

128 *As pediatric infectious disease specialist Dr. Peter Hotez has said*: Eamon N. Dreisbach, "Climate Change: A Growing Threat to Children's Health," *Infectious Diseases in Children 33*: 3 (2020): 1, 10–12.

The Most Dangerous Animal

128 *That he and his wife and daughter had joined*: Hector Caraballo and Kevin King, "Emergency Department Management of Mosquito-Borne Illness: Malaria, Dengue, and West Nile Virus," *Emergency Medicine Practice 16*: 5 (2014): 1–23.

128 *The pathogens it carries*: Winegard, *The Mosquito: A Human History of Our Deadliest Predator*, p. 1.

128 *Most of them were babies and toddlers*: Malaria accounts for most of the historical toll of mosquitos, and the vast majority of those it kills are babies and toddlers: John Whitfield, "Portrait of a Serial Killer," *Nature 3* (2002).

129 *Though mosquito-control programs*: Winegard, *The Mosquito: A Human History of Our Deadliest Predator*, pp. 1–2.

129 *Roughly half of today's mosquito-linked deaths*: The World Health Organization reported 625,000 malaria deaths in 2020, 76 percent of which were children under five, equaling 475,000 children. This is equivalent to 57 percent of the 830,000 global deaths reported in 2018 from all mosquito-borne illnesses (the 2020 total of

global mosquito-illness deaths is not available): WHO, *World Malaria Report 2022* (Geneva: World Health Organization, 2022), p. xxi, accessed July 2, 2023, https://www.who.int/teams/global-malaria-programme/reports/world-malaria-report-2022; Winegard, *The Mosquito: A Human History of Our Deadliest Predator*, p. 1.

129 *Malaria has also been responsible*: Whitfield, "Portrait of a Serial Killer"; Winegard, *The Mosquito: A Human History of Our Deadliest Predator*, pp. 18, 24.

129 *All three groups live in the US*: CDC, "Mosquitos in the United States," *cdc.gov*, accessed July 2, 2023, https://www.cdc.gov/mosquitoes/about/mosquitoes-in-the-us.html.

129 *Outside of sub-Saharan Africa*: Max Roser and Hannah Ritchie, "Malaria," *ourworldindata.org*, accessed July 2, 2023, https://ourworldindata.org/malaria; WHO, *The World Health Report 1999: Making a Difference* (Geneva: World Health Organization, 1999), box 4.2, p. 50, accessed July 2, 2023, https://apps.who.int/iris/handle/10665/42167.

130 *Some parts of the world are becoming too hot or dry*: Colin J. Carlson, Ellen Bannon, Emily Mendenhall et al., "Rapid Range Shifts in African *Anopheles* Mosquitoes over the Last Century," *Biology Letters 19*: 2 (2023): 20220365; Vittor, "Mosquitos, Ticks, and Climate Change."

130 *But for now, the "tropical illness zone"*: NOAA, "Tropical Widening," *sos.noaa.gov*, accessed July 2, 2023, https://sos.noaa.gov/catalog/datasets/tropical-widening; Craig Welch, "Climate Change Pushing Tropical Diseases Toward Arctic," *nationalgeographic.com*, June 14, 2017, accessed July 2, 2023, https://www.nationalgeographic.com/science/article/vibrio-zika-west-nile-malaria-diseases-spreading-climate-change.

130 *In Florida alone*: Greg Allen, "Scientists Find New Invasive Mosquito Species in Florida," *NPR Morning Edition*, March 16, 2021, accessed July 2, 2023, https://www.npr.org/2021/03/16/976598336/scientists-find-new-invasive-mosquito-species-in-florida.

130 *As Dr. Vittor noted*: Micah B. Hahn, Lars Eisen, Janet McAllister et al., "Updated Reported Distribution of *Aedes* (*Stegomyia*) *aegypti* and *Aedes* (*Stegomyia*) *albopictus* (Diptera: Culicidae) in the United States, 1995–2016," *Journal of Medical Entomology 54*: 5 (2017): 1420–1424.

Breakbone Fever

131 *Dengue would infect more than 3 million*: PAHO, "Cases of Dengue in the Americas Exceeded 3 Million in 2019," *paho.org*, February 12, 2020, https://www3.paho.org/hq/index.php?option=com_content&view=article&id=15722:cases-of-dengue-in-the-americas-exceeded-3-million-in-2019&Itemid=0&lang=en#gsc.tab=0.

131 *The small country suffered its worst dengue outbreak*: Lise Alves, "Honduras's Worst Dengue Outbreak in 50 Years," *Lancet 394*: 10196 (2019): 371.

132 *The disease has become thirtyfold more common*: Kristie L. Ebi and Joshua Nealon, "Dengue in a Changing Climate," *Environmental Research 151* (2016): 115–123; Congcong Guo, Zixing Zhou, Zihao Wen et al., "Global Epidemiology of Dengue Outbreaks in 1990–2015: A Systematic Review and Meta-analysis," *Frontiers in Cellular and Infection Microbiology 7* (2017): 317.

132 *Dengue now strikes more than 400 million people a year*: CDC, "Dengue: Data and Maps," *cdc.gov*, last reviewed January 19, 2023, accessed September 10, 2023, https://www.cdc.gov/dengue/statistics-maps/data-and-maps.html.

132 *Historically, almost 95 percent of global dengue cases*: Duane J. Gubler, "Dengue/ Dengue Haemorrhagic Fever: History and Current Status," in *New Treatment Strategies for Dengue and Other Flaviviral Diseases: Novartis Foundation Symposium 277* (Chichester, UK: John Wiley & Sons, Ltd., 2006), pp. 3–22; María G. Guzmán, Gustavo Kouri, Jose Bravo et al., "Effect of Age on Outcome of Secondary Dengue 2 Infections," *International Journal of Infectious Diseases 6*: 2 (2002): 118–124; Lilly M. Verhagen and Ronald de Groot, "Dengue in Children," *Journal of Infection 69* (2014): S77–S86.

132 *The virus was largely wiped from the Americas*: Olivia Brathwaite Dick, Jose L. San Martín, Romeo H. Montoya et al., "The History of Dengue Outbreaks in the Americas," *American Journal of Tropical Medicine and Hygiene 87*: 4 (2012): 584–593.

132 *Over 5,000 people carried dengue into the country*: Aidsa Rivera, Laura E. Adams, Tyler M. Sharp et al., "Travel-Associated and Locally Acquired Dengue Cases—United States, 2010–2017," *Morbidity and Mortality Weekly Report 69*: 6 (2020): 149; Mary Elizabeth Wilson, "Section 11, Posttravel Evaluation: Fever in the Returned Traveler," *CDC Yellow Book 2024* (2023), https://wwwnc.cdc.gov/travel/yellowbook/2024/posttravel-evaluation/fever-in-the-returned-traveler.

132 *If these travelers go home to a community*: Rivera et al., "Travel-Associated and Locally Acquired Dengue Cases—United States, 2010–2017."

132 *Like Zika, the 2019 dengue outbreak*: Tyler M. Sharp, Stephen Morris, Andrea Morrison et al., "Fatal Dengue Acquired in Florida," *New England Journal of Medicine 384*: 23 (2021): 2257–2259.

132 *Scientists who have modeled the impact of climate change*: Jane P. Messina, Oliver J. Brady, Nick Golding et al., "The Current and Future Global Distribution and Population at Risk of Dengue," *Nature Microbiology 4*: 9 (2019): 1508–1515.

132 *Dengue infection raises the odds*: Enny S. Paixão, Maria da Conceição N. Costa, Maria Glória Teixeira et al., "Symptomatic Dengue Infection During Pregnancy and the Risk of Stillbirth in Brazil, 2006–12: A Matched Case-Control Study," *Lancet Infectious Diseases 17*: 9 (2017): 957–964; Enny S. Paixão, Katie Harron, Oona Campbell et al., "Dengue in Pregnancy and Maternal Mortality: A Cohort Analysis Using Routine Data," *Scientific Reports 8*: 9938 (2018): 1–6.

133 *If she is infected with the virus at the time of birth*: Célia Basurko, Severine Matheus, Hélène Hildéral et al., "Estimating the Risk of Vertical Transmission of Dengue: A Prospective Study," *American Journal of Tropical Medicine and Hygiene 98*: 6 (2018): 1826–1832.

133 *About half the newborns who catch chikungunya*: Patrick Gérardin, Georges Barau, Alain Michault et al., "Multidisciplinary Prospective Study of Mother-to-Child Chikungunya Virus Infections on the Island of La Reunion," *PLoS Medicine 5*: 3 (2008): e60.

133 *A breakthrough came in mid-2021*: Joshua M. Wong, Laura E. Adams, Anna P. Durbin et al., "Dengue: A Growing Problem with New Interventions," *Pediatrics 149*: 6 (2022): e2021055522.

133 *No vaccine is yet available*: Guzmán et al., "Effect of Age on Outcome of Secondary Dengue 2 Infections."

The Path to Claudia's Door

134 *Zika erupted from northeastern Brazil*: Sheila S. Barros Brito, Ana Paula M. A. Cunha, C. C. Cunningham et al., "Frequency, Duration and Severity of Drought in the Semiarid Northeast Brazil Region," *International Journal of Climatology 38*: 2 (2018): 517–529; "Brazil's Most Populous Region Facing Worst Drought in 80 Years," *BBC News*, January 24, 2015, accessed June 30, 2023, https://www.bbc.com/news/world-latin-america-30962813.

134 *In October 2015, health officials*: PAHO, "Timeline—Emergence of the Zika Virus in the Americas," *paho.org*, accessed June 30, 2023, https://www.paho.org/en/timeline-emergence-zika-virus-americas.

134 *They found that many of the babies*: Katie Worth and Michelle Mizner, "Zika Uncontained," *Frontline*, August 10, 2016, accessed July 2, 2023, https://www.pbs.org/wgbh/frontline/interactive/zika-water/.

134 *Later analysis of the virus's genes*: Nuno Rodrigues Faria, Raimunda do Socorro da Silva Azevedo, Moritz UG Kraemer et al., "Zika Virus in the Americas: Early Epidemiological and Genetic Findings," *Science 352*: 6283 (2016): 345–349.

135 *The years of the Zika pandemic*: NOAA, National Centers for Environmental Information, *Annual 2016 Global Climate Report*, accessed July 1, 2023, https://www.ncei.noaa.gov/access/monitoring/monthly-report/global/201613.

135 *At these higher temperatures*: Nik Abdull Halim, Nik Muhammad Hanif, Che Dom et al., "A Systematic Review and Meta-analysis of the Effects of Temperature on the Development and Survival of the *Aedes* Mosquito," *Frontiers in Public Health 10* (2022): 1074028.

135 *The insect also became hyperactive*: Rachel Bellone and Anna-Bella Failloux, "The Role of Temperature in Shaping Mosquito-Borne Viruses Transmission," *Frontiers in Microbiology 11* (2020): 584846; Joanna M. Reinhold, Claudio R. Lazzari, and Chloé Lahondère, "Effects of the Environmental Temperature on *Aedes aegypti* and *Aedes albopictus* Mosquitoes: A Review," *Insects 9*: 4 (2018): 158.

135 *Once inside the mosquito*: Blanka Tesla, Leah R. Demakovsky, Erin A. Mordecai et al., "Temperature Drives Zika Virus Transmission: Evidence from Empirical and Mathematical Models," *Proceedings of the Royal Society B: Biological Sciences 285*: 1884 (2018): 20180795.

135 *They were fueled by global warming directly*: Cyril Caminade, Joanne Turner, Soeren Metelmann et al., "Global Risk Model for Vector-Borne Transmission of Zika Virus Reveals the Role of El Niño 2015," *Proceedings of the National Academy of Sciences 114*: 1 (2017): 119–124; Shlomit Paz and Jan C. Semenza, "El Niño and Climate Change—Contributing Factors in the Dispersal of Zika Virus in the Americas?," *Lancet 387*: 10020 (2016): 745.

135 *Severe El Niños occur more often now*: Wenju Cai, Benjamin Ng, Tao Geng et al., "Anthropogenic Impacts on Twentieth-Century ENSO Variability Changes," *Nature Reviews Earth & Environment 4* (2023): 407–418.

136 *That's what happened when the dengue virus returned*: Rachel Lowe, Christovam Barcellos, Patrícia Brasil et al., "The Zika Virus Epidemic in Brazil: From Discovery

to Future Implications," *International Journal of Environmental Research and Public Health 15*: 1 (2018): 96.

136 *Deforestation is a major contributor to global warming*: Raymond E. Gullison, Peter C. Frumhoff, Josep G. Canadell et al., "Tropical Forests and Climate Policy," *Science 316*: 5827 (2007): 985–986.

136 *Between 1990 and 2016 it caused*: Serge Morand and Claire Lajaunie, "Outbreaks of Vector-Borne and Zoonotic Diseases Are Associated with Changes in Forest Cover and Oil Palm Expansion at Global Scale," *Frontiers in Veterinary Science 8* (2021): 230.

136 *It turns out that when lush*: Rory Gibb, David W. Redding, Kai Qing Chin et al., "Zoonotic Host Diversity Increases in Human-Dominated Ecosystems," *Nature 584*: 7821 (2020): 398–402; Nathan D. Burkett-Cadena and Amy Y. Vittor, "Deforestation and Vector-Borne Disease: Forest Conversion Favors Important Mosquito Vectors of Human Pathogens," *Basic and Applied Ecology 26* (2018): 101–110; Chelsea L. Wood, Kevin D. Lafferty, Giulio DeLeo et al., "Does Biodiversity Protect Humans Against Infectious Disease?," *Ecology 95*: 4 (2014): 817–832.

136 *By February 2016*: "Honduras, Nicaragua Report First Cases of Zika in Pregnant Women," *Reuters*, February 4, 2016, accessed June 30, 2023, https://www.reuters.com/article/uk-health-zika-nicaragua/honduras-nicaragua-report-first-cases-of-zika-in-pregnant-women-idUKKCN0VD2OC.

136 *That same month, WHO investigators*: WHO, "WHO Director-General Summarizes the Outcome of the Emergency Committee Regarding Clusters of Microcephaly and Guillain-Barré Syndrome," *who.int*, February 1, 2016, accessed June 30, 2023, https://www.who.int/news/item/01-02-2016-who-director-general-summarizes-the-outcome-of-the-emergency-committee-regarding-clusters-of-microcephaly-and-guillain-barré-syndrome.

Camron

137 *Named after a Connecticut town*: CDC, "How Many People Get Lyme Disease?," *cdc.gov*, accessed July 2, 2023, https://www.cdc.gov/lyme/stats/humancases.html.

138 *Solon is in Johnson County*: Orlan Love, "Spring in Iowa Means—It's Tick Time," *Gazette (IA)*, April 11, 2016, accessed July 2, 2023, https://www.thegazette.com/life/spring-in-iowa-means-its-tick-time/; "Yearly Breakdown of Lyme Disease Cases in Johnson County, Iowa," *TickCheck*, accessed July 2, 2023, https://www.tickcheck.com/stats/county/iowa/johnson-county/lyme; US Census Bureau, QuickFacts: Johnson County, Iowa, *census.gov*, accessed July 6, 2023, https://www.census.gov/quickfacts/johnsoncountyiowa.

138 *But it usually acquires* Borrelia: Brian F. Allan, Felicia Keesing, and Richard S. Ostfeld, "Effect of Forest Fragmentation on Lyme Disease Risk," *Conservation Biology 17*: 1 (2003): 267–272.

139 *Roughly a quarter of Lyme patients*: Amy M. Schwartz, Alison F. Hinckley, Paul S. Mead et al., "Surveillance for Lyme Disease—United States, 2008–2015," *Morbidity and Mortality Weekly Report, Surveillance Summaries 66*: 22 (2017): 1–12.

139 *While he tired easily for several months*: Maureen Monaghan, Stephanie Norman,

Marcin Gierdalski et al., "Pediatric Lyme Disease: Systematic Assessment of Post-treatment Symptoms and Quality of Life," *Pediatric Research* (2023): 1–8.

139 *The most common late manifestation*: Linda K. Bockenstedt and Gary P. Wormser, "Unraveling Lyme Disease," *Arthritis & Rheumatology 66*: 9 (2014): 2313–2323.

139 *The first medical journal article*: Allen C. Steere, Stephen E. Malawista, David R. Snydman et al., "Lyme Arthritis: An Epidemic of Oligoarticular Arthritis in Children and Adults in Three Connecticut Communities," *Arthritis & Rheumatology 20*: 1 (1977): 7–17.

140 *Across the country today*: Kiersten J. Kugeler, Paul S. Mead, Amy M. Schwartz, and Alison F. Hinckley, "Changing Trends in Age and Sex Distributions of Lyme Disease—United States, 1992–2016," *Public Health Reports 137*: 4 (2022): 655–659.

140 *Both the ticks and mice are expanding*: Julie A. Simon, Robby R. Marrotte, Nathalie Desrosiers et al., "Climate Change and Habitat Fragmentation Drive the Occurrence of *Borrelia burgdorferi*, the Agent of Lyme Disease, at the Northeastern Limit of Its Distribution," *Evolutionary Applications 7*: 7 (2014): 750–764.

140 *Peak "tick season" starts earlier*: Patricia A. Nuttall, "Climate Change Impacts on Ticks and Tick-Borne Infections," *Biologia 77* (2022): 1503–1512; Taal Levi, Felicia Keesing, Kelly Oggenfuss, and Richard S. Ostfeld, "Accelerated Phenology of Blacklegged Ticks Under Climate Warming," *Philosophical Transactions of the Royal Society B: Biological Sciences 370*: 1665 (2015): 20130556; Andrew J. Monaghan, Sean M. Moore, Kevin M. Sampson et al., "Climate Change Influences on the Annual Onset of Lyme Disease in the United States," *Ticks and Tick-Borne Diseases 6*: 5 (2015): 615–622.

140 *All these factors are fueling Lyme disease*: EPA, "Key Findings," *Climate Change and Children's Health and Well-Being in the United States* (2023): EPA 430-R-23-001, *epa.gov*, accessed July 6, 2023, https://www.epa.gov/cira/climate-change-and-childrens-health-and-well-being-united-states-report.

140 *The* Ixodes *tick also carries*: CDC, "Powassan Virus: Clinical Evaluation and Disease," *cdc.gov*, accessed July 6, 2023, https://www.cdc.gov/powassan/clinicallabeval.html.

140 *It and other tick-borne diseases*: Ronald Rosenberg, Nicole P. Lindsey, Marc Fischer et al., "Vital Signs: Trends in Reported Vector-Borne Disease Cases—United States and Territories, 2004–2016," *Morbidity and Mortality Weekly Report 67*: 17 (2018): 496–501.

140 *We don't form long-term immunity to Lyme*: Rebecca A. Elsner, Christine J. Hastey, Kimberly J. Olsen, and Nicole Baumgarth, "Suppression of Long-Lived Humoral Immunity Following *Borrelia burgdorferi* Infection," *PLoS Pathogens 11*: 7 (2015): e1004976.

140 *A vaccine might soon be available*: CDC, "Lyme Disease Vaccine," *cdc.gov*, August 11, 2022, accessed July 6, 2023, https://www.cdc.gov/lyme/prev/vaccine.html.

The Incalculable Toll

141 *Caused by the infamous*: WHO, *World Malaria Report*, p. xix.

141 *While anyone can get the disease*: Ibid., p. xxi.

141 *More than 475,000 babies and toddlers*: WHO reported 625,000 malaria deaths in 2020, 76 percent of which were children under five, equaling 475,000. WHO, *World Malaria Report*, p. xxi.

141 *Yet roughly 1,500 people are hospitalized*: Diana Khuu, Mark L. Eberhard, Benjamin N. Bristow et al., "Malaria-Related Hospitalizations in the United States, 2000–2014," *American Journal of Tropical Medicine and Hygiene 97*: 1 (2017): 213–221.

141 *That appears to be what happened*: CDC, "Locally Acquired Cases of Malaria in Florida, Texas, and Maryland," *cdc.gov*, accessed September 4, 2023, https://www.cdc.gov/malaria/new_info/2023/malaria_florida.html.

141 *While it's hard to imagine now*: Sok Chul Hong, "The Burden of Early Exposure to Malaria in the United States, 1850–1860: Malnutrition and Immune Disorders," *Journal of Economic History 67*: 4 (2007): 1001–1035.

141 *In the eighteenth century*: Peter McCanless, "From Paradise to Hospital," in *Slavery, Disease and Suffering in the Southern Lowcountry* (Cambridge, UK: Cambridge University Press, 2011), p. 18.

141 *The disease remained a seasonal plague*: CDC, "Elimination of Malaria in the United States (1947–1951)," *cdc.gov*, July 23, 2018, accessed July 6, 2023, https://www.cdc.gov/malaria/about/history/elimination_us.html.

142 *In some parts of Africa*: Carlson et al., "Rapid Range Shifts in African *Anopheles* Mosquitoes over the Last Century"; Vittor, "Mosquitos, Ticks, and Climate Change."

142 *Analysts predict that childhood mortality*: Shouro Dasgupta, "Burden of Climate Change on Malaria Mortality," *International Journal of Hygiene and Environmental Health 221*: 5 (2018): 782–791.

142 *The World Health Organization still aims*: WHO, "Eliminating Malaria," *who.int*, accessed July 6, 2023, https://www.who.int/activities/eliminating-malaria.

143 Chukchu, *her father told us*: According to the family we lived with, most Ecuadorians today call the disease by its Spanish (and English) name, malaria, as the Kichwa language is being lost among younger generations.

143 *This is how, child by child*: Christophe Rogier, Patrick Imbert, Adama Tall et al., "Epidemiological and Clinical Aspects of Blackwater Fever Among African Children Suffering Frequent Malaria Attacks," *Transactions of the Royal Society of Tropical Medicine and Hygiene 97*: 2 (2003): 193–197.

144 *In the Americas either* P. vivax *or* P. falciparum: Ariel Bardach, Agustín Ciapponi, Lucila Rey-Ares et al., "Epidemiology of Malaria in Latin America and the Caribbean from 1990 to 2009: Systematic Review and Meta-analysis," *Value in Health Regional Issues 8* (2015): 69–79.

144 *After the Kichwa's ancestors shared their knowledge*: Gabriel Gachelin, Paul Garner, Eliana Ferroni et al., "Evaluating Cinchona Bark and Quinine for Treating and Preventing Malaria," *Journal of the Royal Society of Medicine 110*: 1 (2017): 31–40.

144 *But a problem has arisen*: Thomas E. Wellems and Christopher V. Plowe, "Chloroquine-Resistant Malaria," *Journal of Infectious Diseases 184*: 6 (2001): 770–776.

145 Plasmodium*'s growing resistance*: Stephanie L. Richards, Brian D. Byrd, Michael H. Reiskind, and Avian V. White, "Assessing Insecticide Resistance in Adult Mosquitoes: Perspectives on Current Methods," *Environmental Health Insights 14* (2020): 1178630220952790.

145 *To be clear: drugs and pesticides*: James R. Roberts, Catherine J. Karr, Council on Environmental Health et al., "Pesticide Exposure in Children," *Pediatrics 130*: 6 (2012): e1765–e1788.

145 *But a breakthrough finally came*: Apoorva Mandavilli, "A 'Historic Event': First Malaria Vaccine Approved by WHO," *New York Times*, October 6, 2021, accessed July 3, 2023, https://www.nytimes.com/2021/10/06/health/malaria-vaccine-who.html.

145 *Child deaths on the continent fell*: US Agency for International Development, US President's Malaria Initiative, *15 Years of Fighting Malaria and Saving Lives: Annual Report to Congress 2020* (2021), p. 5, https://stacks.cdc.gov/view/cdc/116656/cdc_116656_DS1.pdf

145 *But progress has flattened in the last several years*: Abdisalan M. Noor and Pedro L. Alonso, "The Message on Malaria is Clear: Progress Has Stalled," *The Lancet 399*: 10337 (2022): 1777.

146 *A hundred miles north of her village*: Miguel San Sebastián and Anna Karin Hurtig, "Oil Exploitation in the Amazon Basin of Ecuador: A Public Health Emergency," *Revista Panamericana de Salud Pública 15*: 3 (2004): 205–211.

Swallowed

147 *"She was lovely"*: Lindy Washburn and Monsy Alvarado, "After Zika: Raising a Baby with Microcephaly," *northjersey.com*, April 12, 2018, accessed July 6, 2023, https://www.northjersey.com/story/news/health/2018/04/12/zika-baby-severe-birth-defects/385116002/.

147 *In November 2020 Honduras*: Natalie Kitroeff, "2 Hurricanes Devastated Central America. Will the Ruin Spur a Migration Wave?," *New York Times*, December 4, 2020, accessed December 18, 2023, https://www.nytimes.com/2020/12/04/world/americas/guatemala-hurricanes-mudslide-migration.html.

147 *A physician working for*: Jason Beaubien, "Even Disaster Veterans Are Stunned by What's Happening in Honduras," *NPR*, December 14, 2020, accessed December 18, 2023, https://www.npr.org/sections/goatsandsoda/2020/12/14/945377248/even-disaster-veterans-are-stunned-by-whats-happening-in-honduras.

148 *A doctor in one hard-hit mountain town*: Julie H. Case, "Two Hurricanes, a Clean Water Crisis, and One Can-Do Kid," *teamrubiconusa.org*, January 28, 2021, accessed December 18, 2023, https://teamrubiconusa.org/blog/two-hurricanes-a-clean-water-crisis-and-one-can-do-kid/.

Zikas of Tomorrow

148 *The Zika virus infected thousands of mothers*: Ashley N. Smoots, Samantha M. Olson, Janet Cragan et al., "Population-Based Surveillance for Birth Defects Potentially Related to Zika Virus Infection—22 States and Territories, January 2016–June 2017," *Morbidity and Mortality Weekly Report 69*: 3 (2020): 67–71; Nicole M. Roth, Megan R. Reynolds, Elizabeth L. Lewis et al., "Zika-Associated Birth Defects Reported in Pregnancies with Laboratory Evidence of Confirmed or Possible Zika Virus Infection—US Zika Pregnancy and Infant Registry, December 1, 2015–March 31, 2018," *Morbidity and Mortality Weekly Report 71*: 3 (2022): 73–79.

149 *Research from South America showed*: Léo Pomar, Manon Vouga, Véronique Lambert

et al., "Maternal–Fetal Transmission and Adverse Perinatal Outcomes in Pregnant Women Infected with Zika Virus: Prospective Cohort Study in French Guiana," *British Medical Journal 363* (2018): k4431.

149 *But Zika would also prove to cause*: Elena Marbán-Castro, Laia J. Vazquez Guillamet, Percy Efrain Pantoja et al., "Neurodevelopment in Normocephalic Children Exposed to Zika Virus in Utero with No Observable Defects at Birth: A Systematic Review with Meta-analysis," *International Journal of Environmental Research and Public Health 19*: 12 (2022): 7319.

149 *A new worry, that babies could be harmed*: Jessica Raper, Zsofia Kovacs-Balint, Maud Mavigner et al., "Long-Term Alterations in Brain and Behavior After Postnatal Zika Virus Infection in Infant Macaques," *Nature Communications 11*: 1 (2020): 2534.

149 *In 2021, Zika flared in India*: WHO, "Zika Virus Disease—India," *who.int*, October 14, 2021, accessed July 6, 2023, https://www.who.int/emergencies/disease-out break-news/item/zika-virus-disease-india.

149 *In 2022, Latin America recorded tens of thousands of cases*: PAHO, "Epidemiological Update for Dengue, Chikungunya, and Zika in 2022," *paho.org*, updated January 6, 2024, accessed January 6, 2024, https://ais.paho.org/ha_viz/arbo/pdf/PAHO%20 Arbo%20Bulletin%202022.pdf

149 *Without rapid cuts in greenhouse gas emissions*: Sadie J. Ryan, Colin J. Carlson, Blanka Tesla et al., "Warming Temperatures Could Expose More Than 1.3 Billion New People to Zika Virus Risk by 2050," *Global Change Biology 27*: 1 (2021): 84–93.

149 *When it returns in force, a vaccine*: Yuchen Wang, Lin Ling, Zilei Zhang, and Alejandro Marin-Lopez, "Current Advances in Zika Vaccine Development," *Vaccines 10*:11 (2022): 1816.

149 *In one such attempt in 2021*: Emily Waltz, "First Genetically Modified Mosquitoes Released in the United States," *Nature 593*: 7858 (2021): 175–176.

149 *Other tools*: Wei Huang, Sibao Wang, and Marcelo Jacobs-Lorena, "Use of Microbiota to Fight Mosquito-Borne Disease," *Frontiers in Genetics 11* (2020): 196; William Steigerwald, "Help NASA Track and Predict Mosquito-Borne Disease Outbreaks," *nasa.gov*, July 2, 2018, accessed July 6, 2023, https://www.nasa.gov/feature/god dard/2018/help-nasa-track-and-predict-mosquito-borne-disease-outbreaks; CDC, "Mosquito Surveillance Software," *cdc.gov*, May 16, 2022, accessed July 6, 2023, https://www.cdc.gov/mosquitoes/mosquito-control/professionals/MosqSurvSoft .html.

Chapter Five: The Possible World

Where We Stand

153 *By that year humanity must cut its carbon dioxide (CO_2) emissions*: IPCC, *Global Warming of 1.5°C*; Michael E. Mann, personal communication, December 12, 2023.

153 *That "45 percent by 2030" goal comes from*: IPCC, *Global Warming of 1.5°C*; Jonathan Watts, "We Have 12 Years to Limit Climate Change Catastrophe, Warns UN," *Guardian*, October 8, 2018, accessed March 19, 2023, https://www.theguardian .com/environment/2018/oct/08/global-warming-must-not-exceed-15c-warns-land

mark-un-report; Christopher J. Rhodes, "Only 12 Years Left to Readjust for the 1.5-Degree Climate Change Option—Says International Panel on Climate Change Report: Current Commentary," *Science Progress 102*: 1 (2019): 73–87.

153 *A 45 percent cut from 2010 levels would drop CO_2 emissions*: The 17.9 Gt goal was calculated using total 2010 CO_2 emissions of 32.6 Gt [32.6 – (32.6 x 0.45) = 17.9] as reported in: International Energy Agency (IEA), "Global CO2 Emissions from Energy Combustion and Industrial Processes, 1900–2022," *iea.org*, last updated March 2, 2023, accessed March 19, 2023, https://www.iea.org/data-and-statistics/charts/global-co2-emissions-from-energy-combustion-and-industrial-processes-1900-2022.

153 *But in 2023 those emissions climbed*: Ajit Niranjan, "Global Carbon Emissions from Fossil Fuels to Hit Record High," *Guardian*, December 5, 2023, accessed January 7, 2024, https://www.theguardian.com/environment/2023/dec/05/global-carbon-emissions-fossil-fuels-record.

153 *Methane, the second-most important greenhouse gas*: NOAA, Global Monitoring Laboratory, "Trends in CH4," *gml.noaa.gov*, accessed February 4, 2024, https://gml.noaa.gov/ccgg/trends_ch4/.

153 *This wrong-way trend means*: Piers Forster, Debbie Rosen, Robin Lamboll, and Joeri Rogelj, "Guest Post: What the Tiny Remaining 1.5C Carbon Budget Means for Climate Policy," *CarbonBrief*, November 11, 2022, accessed March 23, 2023, https://www.carbonbrief.org/guest-post-what-the-tiny-remaining-1-5c-carbon-budget-means-for-climate-policy/.

154 *By 2023 we'd used at least a quarter of that*: Ibid.

154 *Although America has contributed more CO_2*: EPA, *Inventory of US Greenhouse Gas Emissions and Sinks: 1990–2021*, *epa.gov*, last updated June 30, 2023, accessed July 6, 2023, https://www.epa.gov/ghgemissions/inventory-us-greenhouse-gas-emissions-and-sinks-1990-2021.

154 *Here the aim is to cut emissions*: US Department of State, US Executive Office of the President, "The Long-Term Strategy of the United States: Pathways to Net-Zero Greenhouse Gas Emissions by 2050," *whitehouse.gov*, November 2021, accessed March 19, 2023, https://www.whitehouse.gov/wp-content/uploads/2021/10/US-Long-Term-Strategy.pdf.

154 *The programs funded by the federal Inflation Reduction Act*: US Department of Energy (DOE), "The Inflation Reduction Act Drives Significant Emissions Reductions and Positions America to Reach Our Climate Goals," *energy.gov*, August 2022, accessed March 19, 2023, https://www.energy.gov/sites/default/files/2022-08/8.18%20InflationReductionAct_Factsheet_Final.pdf.

154 *That's still not enough on our part*: Hannah Ritchie and Max Roser, "CO_2 Emissions," *ourworldindata.org*, accessed March 19, 2023, https://ourworldindata.org/co2-emissions.

154 *The price of solar and wind energy*: Silvio Marcacci, "Renewable Energy Prices Hit Record Lows: How Can Utilities Benefit from Unstoppable Solar and Wind?," *Forbes*, January 21, 2020, accessed March 19, 2023, https://www.forbes.com/sites/energyinnovation/2020/01/21/renewable-energy-prices-hit-record-lows-how-can-utilities-benefit-from-unstoppable-solar-and-wind/; Hannah Ritchie, "The Price of Batteries Has Declined by 97% in the Last Three Decades," *ourworldin*

data.org, June 4, 2021, accessed March 19, 2023, https://ourworldindata.org/battery-price-decline.

154 *And although the world's carbon dioxide emissions*: US Energy Information Administration, "Short-Term Energy Outlook," *eia.gov*, last updated December 12, 2023, accessed January 7, 2024, https://www.eia.gov/outlooks/steo/index.php.

154 *The concentrations of those pollutants in the atmosphere*: NOAA, Global Monitoring Laboratory, "Trends in CO_2, Global Mean Monthly CO_2," *gml.noaa.gov*, accessed January 7, 2024, https://gml.noaa.gov/ccgg/trends/global.html; NOAA, "Carbon Dioxide Now More Than 50% Higher Than Pre-industrial Levels," *noaa.gov*, June 3, 2022, accessed January 7, 2024, https://www.noaa.gov/news-release/carbon-dioxide-now-more-than-50-higher-than-pre-industrial-levels.

154 *Scientists estimate that the world now has only a 5 percent chance*: Peiran R. Liu and Adrian E. Raftery, "Country-Based Rate of Emissions Reductions Should Increase by 80% Beyond Nationally Determined Contributions to Meet the 2 °C Target," *Communications Earth & Environment 2*: 1 (2021): 29.

One Child in the Decades Ahead

155 *But in 2022 a group of Stanford University researchers*: Mark Z. Jacobson, Anna-Katharina von Krauland, Stephen J. Coughlin et al., "Low-Cost Solutions to Global Warming, Air Pollution, and Energy Insecurity for 145 Countries," *Energy & Environmental Science 15* (2022): 3343.

155 *The IPCC recently published a temperature curve*: "Chapter 1: Framing and Context," in IPCC, *Global Warming of 1.5°C*, especially figure 1.4 on p. 62.

156 *As we saw in 2023*: Zeke Hausfather, "State of the Climate: 2023 Smashes Records for Surface Temperature and Ocean Heat."

156 *Of course, if we don't cut fossil fuel use quickly*: Michael E. Mann, personal communication, December 12, 2023.

156 *The latest climate science shows*: Hertsgaard et al., "How a Little-Discussed Revision of Climate Science Could Help Avert Doom."

157 *If air conditioners are still in use*: EPA, "Emissions of Fluorinated Gases," in *Inventory of US Greenhouse Gas Emissions and Sinks: 1990–2021*.

157 *It will likely be richer in plants*: Hana Kahleova, Susan Levin, and Neal Barnard, "Cardio-Metabolic Benefits of Plant-Based Diets," *Nutrients 9*: 8 (2017): 848.

157 *Lower-meat diets*: Joan Sabaté and Sam Soret, "Sustainability of Plant-Based Diets: Back to the Future," *American Journal of Clinical Nutrition 100*, Supplement_1 (2014): 476S–482S; Alyssa Marchese and Alice Hovorka, "Zoonoses Transfer, Factory Farms and Unsustainable Human-Animal Relations," *Sustainability 14*: 9 (2022): 12806.

157 *They will also reduce the greenhouse gas emissions*: Tara Garnett, "Livestock-Related Greenhouse Gas Emissions: Impacts and Options for Policy Makers," *Environmental Science & Policy 12*: 4 (2009): 491–503; Philip G. Curtis, Christy M. Slay, Nancy L. Harris et al., "Classifying Drivers of Global Forest Loss," *Science 361*: 6407 (2018): 1108–1111.

Earthquake

159 *If that seems far-fetched*: Walter J. Daly, "The Black Cholera Comes to the Central Valley of America in the 19th Century—1832, 1849, and Later," *Transactions of the American Clinical and Climatological Association 119* (2008): 143.

159 *In fact, the potential health impacts of the "green revolution"*: Anthony Costello, Mustafa Abbas, Adriana Allen et al., "Managing the Health Effects of Climate Change: *Lancet* and University College London Institute for Global Health Commission," *Lancet 373*: 9676 (2009): 1693–1733; Nick Watts, W. Neil Adger, Paolo Agnolucci et al., "Health and Climate Change: Policy Responses to Protect Public Health," *Lancet 386*: 10006 (2015): 1861–1914.

159 *Because even if fossil fuels were not threatening us*: Duke Nicholas School of the Environment, "Duke Scientist Tells Congress Curbing Global Warming Will Prevent 4.5 Million US Deaths," *nicholas.duke.edu*, August 5, 2020, accessed March 23, 2023, https://nicholas.duke.edu/news/duke-scientist-tells-congress-curbing-global-warming-will-prevent-45-million-us-deaths.

159 *That's largely because of the recent passage*: DOE, "The Inflation Reduction Act Drives Significant Emissions Reductions and Positions America to Reach Our Climate Goals."

160 *And in the spirit of New Deal programs*: National Sustainable Agriculture Coalition, "Inflation Reduction Act of 2022: A Deep Dive on an Historic Investment in Climate and Conservation Agriculture," *sustainableagriculture.net*, August 19, 2022, accessed March 23, 2023, https://sustainableagriculture.net/blog/inflation-reduction-act-of-2022-a-deep-dive-on-an-historic-investment-in-climate-and-conservation-agriculture/.

160 *There, in 2016, Tesla*: Kirsten Korosec, "Tesla's Elon Musk Unveils a Factory He Hopes Will Change the World," *Fortune*, July 30, 2016, accessed March 23, 2023, https://fortune.com/2016/07/30/tesla-gigafactory-elon-event/.

160 *In response to Tesla's success*: Paul Lienert and Ben Klayman, "Ford Follows GM, VW with Two New Dedicated EV Platforms by 2025—Sources," *Reuters*, May 25, 2021, accessed March 23, 2023, https://www.reuters.com/business/sustainable-business/exclusive-ford-follows-gm-vw-with-two-new-dedicated-ev-platforms-by-2025-sources-2021-05-25/.

160 *Meanwhile, more than a dozen states*: Center for Climate and Energy Solutions, "US State Clean Vehicle Policies and Incentives," *c2es.org*, August 2022, accessed March 23, 2023, https://www.c2es.org/document/us-state-clean-vehicle-policies-and-incentives/; EPA, "Regulations for Greenhouse Gas Emissions from Passenger Cars and Trucks," *epa.gov*, updated April 12, 2023, accessed January 7, 2024, https://www.epa.gov/regulations-emissions-vehicles-and-engines/regulations-greenhouse-gas-emissions-passenger-cars-and.

160 *So while these cars were expected to account for*: Neil King, "EVs Forecast to Account for Two Thirds of Global Light-Vehicle Sales in 2035," *ev-volumes.com*, accessed January 12, 2024, https://www.ev-volumes.com.

160 *At the same time, many local utilities now allow*: DOE, "Buying Clean Energy," *energy.gov*, accessed March 23, 2023, https://www.energy.gov/energysaver/buying-clean-electricity.

161 *Though these steps have been financially out of reach*: The White House, "Investing in America—CleanEnergy.gov," *whitehouse.gov*, accessed March 24, 2023, https://www.whitehouse.gov/cleanenergy/.

161 *There are many other actions parents can take*: Paul Hawken, ed., *Drawdown: The Most Comprehensive Plan Ever Proposed to Reverse Global Warming* (London, UK: Penguin, 2017).

161 *Climate-helping projects for children*: Mary DeMocker, *The Parents' Guide to Climate Revolution: 100 Ways to Build a Fossil-Free Future, Raise Empowered Kids, and Still Get a Good Night's Sleep* (Novato, CA: New World Library, 2018).

161 *They rightly point out that only one hundred companies*: Tess Riley, "Just 100 Companies Responsible for 71% of Global Emissions, Study Says," *Guardian*, July 10, 2017, accessed March 24, 2023, https://www.theguardian.com/sustainable-business/2017/jul/10/100-fossil-fuel-companies-investors-responsible-71-global-emissions-cdp-study-climate-change; Mark Kaufman, "The Carbon Footprint Sham," *mashable.com*, accessed July 3, 2023, https://mashable.com/feature/carbon-footprint-pr-campaign-sham.

161 *And, as the IRA underscores*: Fred Lambert, "Complete Breakdown of the $4.9 Billion in Government Support the *LA Times* Claims Elon Musk's Companies Are Receiving," June 5, 2015, accessed March 24, 2023, https://electrek.co/2015/06/02/complete-breakdown-of-the-4-9-billion-in-government-support-the-la-times-claims-elon-musks-companies-are-receiving/; EPA, "Policies and Regulations," *epa.gov*, accessed March 24, 2023, https://www.epa.gov/green-power-markets/policies-and-regulations.

Fighting Back

162 *Anger, like hope, is an energizing emotion*: Studies have found that "constructive hope," or the belief that action can make a difference, is motivating, while optimism, or "passive hope," is not: Maria Ojala, "Hope and Climate Change: The Importance of Hope for Environmental Engagement Among Young People," *Environmental Education Research 18*: 5 (2012): 625–642; Howard Frumkin, "Hope, Health, and the Climate Crisis," *Journal of Climate Change and Health 5* (2022): 100115; Julia Sangervo, Kirsti M. Jylhä, and Panu Pihkala, "Climate Anxiety: Conceptual Considerations, and Connections with Climate Hope and Action," *Global Environmental Change 76* (2022): 102569; Maria Ojala, "Hope and Climate-Change Engagement from a Psychological Perspective," *Current Opinion in Psychology 49* (2023): 101514.

162 *Young people have also taken their protest to the courts*: *Juliana v. United States*, No. 6:15-cv-01517-TC (Oregon District Court, Eugene Division, 2015), accessed July 5, 2023, https://casetext.com/case/juliana-v-united-states-2.

162 *Similar lawsuits have been filed at state and international levels*: Our Children's Trust, "State Legal Actions" and "Global Legal Actions," *ourchildrenstrust.org*, accessed March 24, 2023, https://www.ourchildrenstrust.org/state-legal-actions and https://www.ourchildrenstrust.org/global-legal-actions.

162 *In* Juliana, *for example, the Ninth Circuit ruled*: John Schwartz, "Court Quashes Youth Climate Change Case Against Government," *New York Times*, January 17,

2020, accessed March 24, 2023, https://www.nytimes.com/2020/01/17/climate/juliana-climate-case.html.

162 *A breakthrough came in August 2023*: Sam Bookman, "Held v. Montana: A Win for Young Climate Advocates and What It Means for Future Litigation," Environment and Energy Law Program, Harvard Law School, *eelp.law.harvard.edu*, August 30, 2023, accessed September 4, 2023, https://eelp.law.harvard.edu/2023/08/held-v-montana/.

162 *Some of the children testified that*: Amy Beth Hanson and Matthew Brown, "Young Athlete in Montana Climate Change Trial Testifies He Uses Inhaler Due to Forest Fire Smoke," *apnews.com*, June 13, 2023, accessed September 4, 2023, https://apnews.com/article/youth-climate-trial-montana-bee4111fe38e2cdda7c1cb4177cd558f.

162 *That same month, young and indigenous activists in Ecuador*: Manuela Andreoni and David Gelles, "Ecuador Will Keep Some Oil in the Ground," *New York Times*, August 22, 2023, accessed September 5, 2023, https://www.nytimes.com/2023/08/22/climate/ecuador-will-keep-some-oil-in-the-ground.html.

163 *While they chanted in the streets*: Global Witness, "Hundreds of Fossil Fuel Lobbyists Flooding COP26 Climate Talks," *globalwitness.org*, November 8, 2021, accessed March 25, 2023, https://www.globalwitness.org/en/press-releases/hundreds-fossil-fuel-lobbyists-flooding-cop26-climate-talks/.

163 *By the time COP28 was held in Dubai*: Jeffrey Kluger, "There Are More Fossil Fuel Lobbyists Than Ever at COP28," *Time*, December 5, 2023, accessed January 15, 2024, https://time.com/6342799/fossile-fuel-lobbyists-COP28/.

163 *The resulting agreement*: Karl Mathiesen, Zack Colman, Zia Weise et al., "COP28 Ends with First-Ever Call to Move Away from Fossil Fuels," *Politico*, December 12, 2023, updated December 13, 2023, accessed January 15, 2024, https://www.politico.com/news/2023/12/12/newest-cop28-climate-summit-text-00131257.

163 *Climate scientist Michael Mann*: Michael E. Mann and Susan Joy Hassol, "Op-Ed: Glasgow's Hope at a Critical Moment in the Climate Battle," *Los Angeles Times*, November 13, 2021, accessed March 25, 2023, https://www.latimes.com/opinion/story/2021-11-13/cop26-glasgow-climate-change.

"Dear Friends"

163 *Lillian Fortuna was only ten years old*: Lillian Fortuna and Mila Drumke, personal communication, October 11, 2022; Greta Thunberg, *No One Is Too Small to Make a Difference* (London, UK: Penguin, 2019).

165 *Just three days after the attack began*: Bill McKibben, "This Is How We Defeat Putin and Other Petrostate Autocrats," *Guardian*, February 25, 2022, accessed March 25, 2023, https://www.theguardian.com/commentisfree/2022/feb/25/this-is-how-we-defeat-putin-and-other-petrostate-autocrats.

165 *Within weeks her petition had topped*: Lillian Fortuna, "Stop Putin by Sending Heat Pumps to Europe," *change.org*, March 12, 2022, accessed March 25, 2023, https://www.change.org/p/joseph-r-biden-stop-putin-by-sending-heat-pumps-to-europe.

165 *In June 2022 Biden did, in fact*: DOE, "President Biden Invokes Defense Production

Act to Accelerate Domestic Manufacturing of Clean Energy," June 6, 2022, accessed March 25, 2023, https://www.energy.gov/articles/president-biden-invokes-defense-production-act-accelerate-domestic-manufacturing-clean.

165 *"I think it made everyone happy"*: Bill McKibben, personal communication, October 29, 2022.

166 *He had spoken to her through his work*: Third Act, "Lillian Fortuna—Bill Mckibben .mp4," Vimeo video, 03:31, April 27, 2022, accessed March 25, 2023, https://vimeo.com/703896268.

166 *Solnit took a broader view*: Rebecca Solnit, personal communication, October 30, 2022.

166 *"Lots of New York buildings are in the exploratory phase"*: New York State, "The Future of Buildings: New York's Carbon Neutral Buildings Roadmap—Executive Summary," *nyserda.ny.gov*, December 2022, accessed July 6, 2023, https://www.nyserda.ny.gov/All-Programs/Carbon-Neutral-Buildings.

167 *She laughs at herself*: Sengupta, "Protesting Climate Change, Young People Take to Streets in a Global Strike."

Every Pound of Carbon

167 *But just a month after its passage*: Lisa Friedman and Coral Davenport, "Senate Ratifies Pact to Curb a Broad Category of Potent Greenhouse Gases," *New York Times*, September 21, 2022, accessed March 25, 2023, https://www.nytimes.com/2022/09/21/climate/hydrofluorocarbons-hfcs-kigali-amendment.html.

168 *Home to one of the largest economies in the world*: Russ Mitchell, "California Bans Sales of New Gas-Powered Cars by 2035. Now the Real Work Begins," *Los Angeles Times*, August 25, 2022, accessed March 25, 2023, https://www.latimes.com/business/story/2022-08-25/california-ban-gasoline-mandate-zero-emission-2035; Natalie Neysa Alund, "California's Pioneering Climate Change Plans to Nix Gas Heater Sales by 2030," *USA Today*, September 24, 2022, accessed March 25, 2023, https://www.usatoday.com/story/money/energy/2022/09/24/california-bans-natural-gas-furnaces-heaters-2030/8103912001/.

168 *"After so many years of doing next to nothing"*: Robinson Meyer, "The Senate Just Quietly Passed a Major Climate Treaty," *Atlantic*, September 28, 2022, accessed March 25, 2023, https://www.theatlantic.com/science/archive/2022/09/congress-climate-policy-hydrofluorocarbons-kigali-amendment/671579/.

168 *Current international pledges*: United Nations Environmental Programme, *Emissions Gap Report 2023: Broken Record - Temperatures Hit New Highs, Yet World Fails to Cut Emissions (Again)*.

168 *While that's an improvement*: David Wallace-Wells. "Beyond Catastrophe—A New Climate Reality Is Coming into View," *New York Times Magazine*, October 26, 2022, accessed March 25, 2023, https://www.nytimes.com/interactive/2022/10/26/magazine/climate-change-warming-world.html.

168 *"People do not understand the magnitude"*: Fiona Harvey, "Top Scientist: We Can't Adapt Our Way out of This Climate Crisis," *Mother Jones*, June 2, 2022, accessed March 25, 2023, https://www.motherjones.com/environment/2022/06/climate-scientist-katharine-hayhoe-crisis-adaptation-global-warming-impact/.

169 *Many states and municipalities have prepared "climate action plans"*: Center for Climate and Energy Solutions, "US State Climate Action Plans," *c2es.org*, December 2022, accessed March 25, 2023, https://www.c2es.org/document/climate-action-plans/; Georgetown Climate Center, "State Adaptation Progress Tracker," *georgetownclimate .org*, accessed March 25, 2023, https://www.georgetownclimate.org/adaptation /plans.html.

170 *The country's largest banks*: Kat Taylor and Bill McKibben, "Op-Ed: If You Bank with the Big 4, Your Money Has an Alarming Carbon Footprint," *Los Angeles Times*, January 18, 2023, accessed March 25, 2023, https://www.latimes.com/opinion /story/2023-01-18/fossil-fuel-expansion-bank-financed-emissions.

170 *Has your company's retirement fund*: Benjamin A. Jones, Andrew L. Goodkind, and Robert P. Berrens, "Economic Estimation of Bitcoin Mining's Climate Damages Demonstrates Closer Resemblance to Digital Crude Than Digital Gold," *Scientific Reports 12*: 1 (2022): 14512; Jon Huang, Claire O'Neill, Hiroko Tabuchi, "Bitcoin Uses More Electricity Than Many Countries. How Is That Possible?," *New York Times*, September 3, 2021, accessed March 25, 2023, https://www.nytimes.com /interactive/2021/09/03/climate/bitcoin-carbon-footprint-electricity.html.

170 *A nationwide movement to decarbonize schools*: Meredith Deliso, "How Schools Are Combating Climate Change, from Green Schoolyards to Solar Power," November 4, 2021, *abcnews.go.com*, accessed October 16, 2023, https://abcnews.go.com/US /schools-combatting-climate-change-green-schoolyards-solar-power/story?id=8079 9147.

170 *Multiple states are transitioning from diesel school buses*: Lydia Freehafer, Leah Lazer, and Brian Zepka, "The State of Electric School Bus Adoption in the US," World Resources Institute, September 21, 2023, accessed October 16, 2023, https://www .wri.org/insights/where-electric-school-buses-us; EPA, "Clean School Bus Program," *epa.gov*, accessed January 15, 2024, http://www.epa.gov/cleanschoolbus.

170 *Two nonprofit groups*: New Buildings Institute, *Decarbonization Roadmap Guide for School Building Decision Makers, newbuildings.org*, April 25, 2022, accessed October 16, 2023, https://newbuildings.org/resource/decarbonization-roadmap -guide-for-school-building-decision-makers/; World Resources Institute, "Electric School Bus Initiative," *electricschoolbusinitiative.org*, accessed October 16, 2023, https://electricschoolbusinitiative.org.

171 *Vote for leaders who will end fossil fuel subsidies*: Simon Black, Ian Parry, and Nate Vernon, "Fossil Fuel Subsidies Surged to Record $7 Trillion," *imf.org*, August 24, 2023, accessed February 23, 2024, https://www.imf.org/en/Blogs/Articles/2023/08/24 /fossil-fuel-subsidies-surged-to-record-7-trillion.

171 *Stop buying products containing palm oil*: Abrahm Lustgarten, "Palm Oil Was Supposed to Help Save the Planet. Instead it Unleashed a Catastrophe," *New York Times*, November 20, 2018, accessed February 24, 2024, https://www.ny times.com/2018/11/20/magazine/palm-oil-borneo-climate-catastrophe.html.

171 *"We can't give in to despair"*: Katharine Hayhoe, "The Most Important Thing You Can Do to Fight Climate Change: Talk About It," *katharinehayhoe.com*, January 11, 2019, accessed March 25, 2023, http://www.katharinehayhoe.com/2019/01/11/the -most-important-thing-you-can-do-to-fight-climate-change-talk-about-it/.

171 *As Michael Mann detailed*: Mann, *The New Climate War*, "Chapter 4: It's YOUR Fault," pp. 68–72, and "Chapter 6: Sinking the Competition," pp. 123–146.

Ashes to Flowers

172 *In the past, affected boys rarely lived beyond young adulthood*: Dongsheng Duan, Nathalie Goemans, Shin'ichi Takeda et al., "Duchenne Muscular Dystrophy," *Nature Reviews Disease Primers 7*: 1 (2021): 13.

172 *It's my personal remnant of humanity's battle*: "JAMA Revisited: Current Comment," *JAMA 326*: 21 (2021): 2217, a reprint of *JAMA 77*: 23 (1921): 1823.

172 *The disease killed an estimated 300 million people*: Donald A. Henderson, "The Eradication of Smallpox—An Overview of the Past, Present, and Future," *Vaccine 29* (2011): D7-D9.

Index

About the Author

Debra Hendrickson, MD, is a board-certified pediatrician practicing in Reno, Nevada. She is a clinical professor of pediatrics at the University of Nevada School of Medicine, where she lectures on the impact of early childhood experiences (such as poverty and trauma) on long-term health. She has an honors degree in environmental studies from Brown University, and was an environmental analyst and planner in New England and Seattle for ten years before attending medical school. Dr. Hendrickson has received many awards for academic achievement and research in both environmental studies and medicine. She is a fellow of the American Academy of Pediatrics, a member of its Council on Environmental Health and Climate Change, and a founding member of Nevada Clinicians for Climate Action. She has three grown children.